Over Land and Sea

Over Land and Sea

Migration from Antiquity to the Present Day

MASSIMO LIVI-BACCI

Translated by David Broder

polity

First published in Italian as *Per terre e per mari. Quindici migrazioni dall'antichità ai nostri giorni* © 2021 by Società editrice il Mulino, Bologna

This English edition © Polity Press, 2023

This book has been translated thanks to a translation grant awarded by the Italian Ministry of Foreign Affairs and International Cooperation / Questo libro è stato tradotto grazie a un contributo alla traduzione assegnato dal Ministero degli Affari Esteri e della Cooperazione Internazionale italiano.

Polity Press
65 Bridge Street
Cambridge CB2 1UR, UK

Polity Press
111 River Street
Hoboken, NJ 07030, USA

All rights reserved. Except for the quotation of short passages for the purpose of criticism and review, no part of this publication may be reproduced, stored in a retrieval system or transmitted, in any form or by any means, electronic, mechanical, photocopying, recording or otherwise, without the prior permission of the publisher.

ISBN-13: 978-1-5095-5529-1 – hardback
ISBN-13: 978-1-5095-5530-7 – paperback

A catalogue record for this book is available from the British Library.

Library of Congress Control Number: 2022952026

Typeset in 11 on 14pt Warnock Pro
by Cheshire Typesetting Ltd, Cuddington, Cheshire
Printed and bound in Great Britain by CPI Group (UK) Ltd, Croydon

The publisher has used its best endeavours to ensure that the URLs for external websites referred to in this book are correct and active at the time of going to press. However, the publisher has no responsibility for the websites and can make no guarantee that a site will remain live or that the content is or will remain appropriate.

Every effort has been made to trace all copyright holders, but if any have been overlooked the publisher will be pleased to include any necessary credits in any subsequent reprint or edition.

For further information on Polity, visit our website:
politybooks.com

Contents

Illustrations   vii

Introduction   1

I. Antiquity   7
1.1. Seneca, two thousand years ago   7
1.2. Settlers and founders: *ápoikoi* and *oikistés*   12
1.3. Augustus's *Res gestae*   16
1.4. Peoples on the march   23

II. In the Hands of the State   28
2.1. Forced migration   28
2.2. Peru: up and down the Andes   32
2.3. The end of an empire   40
2.4. The Soviet Union and its internal enemies   46

III. Misdeeds of Nature 55
3.1. Unkind nature 55
3.2. Drought 58
3.3. A Caribbean odyssey 63
3.4. Ireland: the blight of diaspora 70

IV. Organized Migration 77
4.1. On the road, not alone 77
4.2. The *filles du roi* in the laboratory of *Nouvelle France* 80
4.3. The *Drang nach Osten* and the Germanization of Eastern Europe 86
4.4. From the Rhine to the Volga with Catherine the Great 93

V. Free Migration 103
5.1. A rare phenomenon 103
5.2. Moving freely 106
5.3. The great transoceanic migration 113
5.4. America: the 'advancing wave' of migration 120

Reconsiderations 128

Notes 134
Index 154

Illustrations

Map A: From where, to where: outline map of the migratory movements within Europe covered in this book (Copyright © Società editrice il Mulino).   5
Map B: Outline map of the migratory movements within the Americas covered in this book (Copyright © Società editrice il Mulino).   6

Plate section
1. Greek colonies in the Mediterranean. © Geo4Map – Novara.
2. Ethnicities of Germania, based on the writings of Pliny (*Naturalis Historia*) and Tacitus (*Germania*), late 1st century AD. Wikimedia Commons.
3. The *Res gestae* of Augustus, fragment of an inscription found in Ankara. Wikimedia Commons.
4. Battle between Goths and Romans, Ludovisi Sarcophagus, circa 251/252 AD, Museo Nazionale Romano, Palazzo Altemps. Wikimedia Commons.
5. The Incas' road system consisted of two parallel routes, one coastal (roughly from the region of Guayaquil to the region of Santiago de Chile)

and one mountainous (from Quito in Ecuador to the region of Mendoza in Argentina), each about 5,000 km long, with many cross-connections. © Emily Carter.
6. The city of Potosí, Bolivia. Taken from John Ogilby, *America, Being the Latest, and Most Accurate Description of the New World*, London, 1671. The Cerro (mountain) of Potosí provided more than half of the silver destined for Europe, with the employment of thousands of migrants (*mitayos*) who arrived from as far as a thousand kilometres away for the annual *mita* (*corvée*). © Album / Alamy Stock Photo.
7. Deportations and population exchanges, Balkan front, Greece and Turkey, 1915–1925.
8. Deportation of Germans from Norka, a colony founded in the Volga region in 1767, 60 kilometres south of Saratov, September 1941. 1.2 million Germans were deported between September 1941 and January 1942.
9. Dust Bowl: the arrival of a dust storm. Wikimedia Commons.
10. Port-au-Prince, capital of Haiti, destroyed by the earthquake on 12 January 2010. © REUTERS / Alamy Stock Photo.
11. Allegory of Ireland, Hunger, Emigration and a 'Coffin Ship'. Thomas Nast, 'The Herald of Relief from America', Library of Congress, 91732265.
12. Germanic *Drang nach Osten:* foundation of a village, under the leadership of a *locator*. Illustration from the 'Sachenspiegel' by Eike von Repgow, 13th century (Cpg. 154, Library of the University of Heidelberg). Between the 11th and 14th centuries, thousands of villages were founded east of the line made up by the rivers Elbe and Saale. Wikimedia Commons.

13. The arrival of the *filles du roi* in Quebec, 1667, painting by Charles William Jefferys, Paris, Bibliothéque Nationale de France. Wikimedia Commons.
14. Portrait of Catherine the Great, at the time of the German migration to the Volga region, attributed to Giovanni Battista Lampi the Elder, c.1793, private collection. Wikimedia Commons.
15. A transhumant shepherd of the Mesta (a powerful sheep breeders' association) in Spain. Millions of sheep transhumed annually, up to a maximum of 5 million at the end of the 18th century, accompanied by many tens of thousands of shepherds. © Gianni Dagli Orti/Shutterstock
16. The meeting of the two branches of the Transcontinental Railroad, built by the Central Pacific and Union Pacific, at Promontory Point, Utah, 10 May 1869. Wikimedia Commons.
17. Irish emigrants set off for the United States, from *The Illustrated London News*, 6 July 1850. Between 1845 and 1850, 1.8 million Irish migrants left their country on the 'Coffin Ships'. Wikimedia Commons.
18. The *Titanic* in Southampton harbour. Having set off 10 April 1912, it sank five days later; there were 2,233 people aboard, of whom 1,503 (67%) perished. First class accommodated 325 rich passengers, of whom 125 (38%) perished, but there were also 706 third-class passengers, almost all British, Irish and Scandinavian migrants, of whom 528 (75%) perished. Wikimedia Commons.
19. Italian migrants head to the Opera assistenza emigranti upon arrival in Buenos Aires. According to official statistics, between 1876 and 1930, almost 2.5 million Italian expatriates reached Argentina. © MARKA/Alamy Stock Photo.

Introduction

Changing abode is a prerogative of human beings who migrate from one location to another, whether to flee from danger or to seek out new opportunities. For hundreds of thousands of years, humankind gradually spread across the planet sustained by this instinct – a quality innate to our species. Even today, in an age of sedentary populations, mobility is a pervasive social phenomenon, functional to our existence. During their lives, almost all our contemporaries have experienced one or more migrations, but they also vary in their distance and duration, and may be driven by the most diverse motives. At the same time, migration as a phenomenon is difficult to define and classify, given the various factors that drive it; the different ways in which it takes place; its intermittent flow over time; and the multiple circumstances surrounding it. Migration is a biodemographic phenomenon, for it implies genetic and ethnic mixing. Migration is a social phenomenon, for it is a factor in the turnover and renewal of communities. Migration is a political fact, for it influences government decisions. Migration eludes generalizations, paradigms or models, even though these do provide tools, or instruments, necessary for ordering and narrating what we do understand.

In the pages of this book, I have knotted together the scattered threads of observations I have made over time on phenomena regarding migration. I have chosen to do this by telling a series of stories – fifteen in all – ranging from Antiquity to the present day, and concerning America and Europe, to the exclusion of the other continents, which are beyond my remit here. These stories do not, and cannot, constitute an embryo of a history of migrations and are not integrated into any systematic, chronological or geographical treatment of my subject. They are, however, interconnected by a criterion that allows us to compare different eras, peoples and contexts. For if the inclination towards mobility is intrinsic to human nature, it is also true that migrations are not necessarily a voluntary act, a matter of free choice, resulting from an individual decision. They may be forced or the result of strong pressure. It is thus worth considering migrations – both past and present – also taking into account what degree of free choice lay behind them.

This freedom may be non-existent, when migration is 'forced', imposed by the overbearing actions of some tyrant or strongman; or it is compelled by a conflict, or a war, that endangers survival; or, again, by a natural disaster, an earthquake, a drought, a virus, a destructive micro-organism. At the opposite end of the spectrum are fully voluntary and deliberate migrations, based on a choice made after weighing up the pros and cons of moving, and where it would also have been possible not to leave. Between these two poles there is a continuum of different situations, with gradually increasing measures of voluntariness and choice. A broad middle segment of this continuum consists of 'organized' migrations: that is, ones that would not have taken place if not for the outside organizational and financial support of a government, a lord or an elite, whose intervention substantially shifts the balance of benefits and disadvantages to be expected from migration.

The fifteen examples of migration here collected span these typologies: they vary considerably, but they are grouped in each chapter with reference to similar assumptions. Thus, among the forced migrations we find the case of the *mitimaes*, colonists transplanted from one side of the Inca Empire to the other during its two centuries of expansion; or the deportations and expulsions of minorities caused by the dissolution of the Ottoman Empire and, two decades later, by the conflicts of the Second World War. At the other extreme, among the cases of free migration – the result of voluntary choices – we find the movement of workers within the Western Europe of the seventeenth and eighteenth centuries, or the great transoceanic migration of the late nineteenth and the early twentieth. The Germanization of Central and Eastern Europe in the late Middle Ages, on the eastern side of the Elbe, took place under the organizing impulse of the great economic potentates that financed it; the migrations that led to the settlement of Greek colonies on the Black Sea and the coasts of the Mediterranean are also an example of organized migration. In North America, the great nineteenth-century migration from the Atlantic to the Pacific was, essentially, a free migration, but one that was intertwined with the forced expulsion of native populations (for instance the Trail of Tears). At the same time, America received migrants driven out of Ireland by a natural disaster, the potato blight, bringing a mass of refugees from the Great Famine. In the following century, another disaster – the drought that caused the Dust Bowl – would set hundreds of thousands of migrants on their way to California. Further north, the arrival of the 'Filles du Roi' – the young girls sent by Colbert and Louis XIV to Quebec – was an example of organized migration for demographic purposes, which though small in numbers surely was the focus of careful preparations.

The fifteen stories are brief forays into the millennia-long history of migration, in a search for paradigms that bind them together. All the cases of forced migration, caused by political violence or natural disasters, generated unspeakable suffering. Organized migrations had positive consequences when some external intervention gave migrants a greater chance of success and strengthened their fitness or capacity to adapt to new conditions. But in many cases the assumptions and calculations behind them were wrong and led to painful failures. Freely chosen migration has generally been successful; it is the desirable form of all consciously made movements, and it is not far-fetched to say that historically the successes have outweighed the failures.

Two last considerations. The first is that – judging from what can be garnered from the history of the last two millennia – humanity's spread across the planet owes in considerable part to non-free migration. We need only think of the Americas, where the population of supposed African origin, descended from the victims of the slave trade – the most total and absolute form of forced migration – stands at about two hundred million out of a total population of one billion. Or of the almost one hundred million refugees and displaced persons registered by international organizations today. Contemporary examples of migration have no place in these pages, except for one instance. They have deliberately been excluded, since contemporary migration flows are well known, studied and recounted. What the reader will be able to do is compare what is happening in today's world with our narration of the often-forgotten movements of eras past.

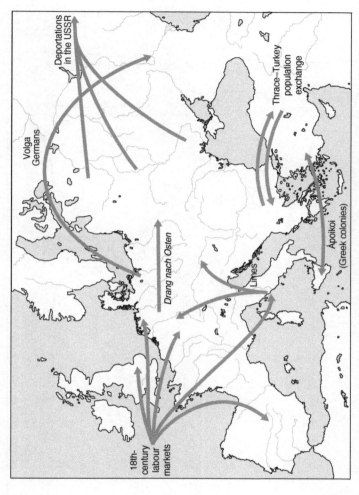

Map A. From where, to where: outline map of the migratory movements within Europe covered in this book (Copyright © Società editrice il Mulino)

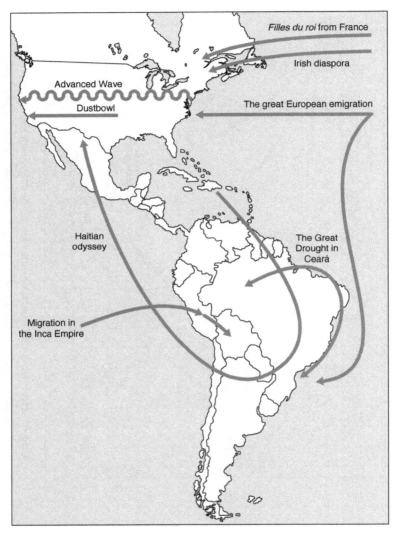

Map B: Outline map of the migratory movements within the Americas covered in this book
(Copyright © Società editrice il Mulino).

I
Antiquity

1.1. Seneca, two thousand years ago

I find some writers who declare that mankind has a natural itch for change of abode and alteration of domicile: for the mind of man is wandering and unquiet; it never stands still, but spreads itself abroad and sends forth its thoughts into all regions, known or unknown; being nomadic, impatient of repose, and loving novelty beyond everything else.[1]

Seneca, exiled to Corsica by the emperor Claudius, wrote these words in his epistle to his mother Helvia. For him, man's nature is not 'formed from the same elements as the heavy and earthly body, but from heavenly spirit: now heavenly things are by their nature always in motion, speeding along and flying with the greatest swiftness.' And the human spirit revels in this, moved as it is by an intimate need for change. The same goes for peoples and nations. Indeed:

What is the meaning of Greek cities in the midst of barbarous districts? or of the Macedonian language existing among the Indians and the Persians? Scythia and all that region which swarms with wild and uncivilized tribes boasts nevertheless Achaean cities along the shores of the Black Sea. Neither the rigours of eternal winter, nor the character of men as savage as their climate, has prevented people migrating thither. There is a mass of Athenians in Asia Minor. Miletus has sent out into various parts of the world citizens enough to populate seventy-five cities. . . . Asia claims the Tuscans as her own: there are Tyrians living in Africa, Carthaginians in Spain; Greeks have pushed in among the Gauls, and Gauls among the Greeks. The Pyrenees have proved no barrier to the Germans.[2]

Two thousand years ago, for Seneca, the known world was a melting pot of many ethnicities, cultures and languages; it was a world of migrants, driven along often impervious and unknown routes by human nature itself:

men drag along with them their children, their wives, and their aged and worn-out parents. Some have been tossed hither and thither by long wanderings, until they have become too wearied to choose an abode, but have settled in whatever place was nearest to them: others have made themselves masters of foreign countries by force of arms: some nations while making for parts unknown have been swallowed up by the sea: some have established themselves in the place in which they were originally stranded by utter destitution.[3]

If human nature is mobile – that is, if humanity is predisposed to migration – there must still be some cause that prompts a concrete move. For migration is nothing but the abandonment of one's living context, one's customs, one's *domus*, one's home:

Nor have all men had the same reasons for leaving their country and for seeking for a new one: some have escaped from their cities when destroyed by hostile armies, and having lost their own lands have been thrust upon those of others: some have been cast out by domestic quarrels: some have been driven forth in consequence of an excess of population, in order to relieve the pressure at home: some have been forced to leave by pestilence, or frequent earthquakes, or some unbearable defects of a barren soil: some have been seduced by the fame of a fertile and over-praised clime. . . . the movement of the human race is perpetual: in this vast world some changes take place daily. The foundations of new cities are laid, new names of nations arise, while the former ones die out, or become absorbed by more powerful ones.[4]

I have prefaced my thoughts with these eloquent passages, written some two millennia ago, because they provide fitting guidance for our efforts to interpret the vicissitudes of human migration. They could serve as an apt beginning for a modern treatise on migrations, if that type of literature still existed. Indeed, in these words we find all the themes of modern debate on this subject. First, the fact that migration is inherent in the human species, as it is to all animal species, just as the stars – and nature – are always 'flying along with the greatest swiftness'. Still today, as in Seneca's time, the mixing of peoples and ethnicities – a consequence of the historical stratification of migratory movements – is a self-evident reality. Today, several hundred million people do not live in the country where they were born; in the Roman context, Seneca had no statistics to back him up, but he could draw on his own observations, the testimony of contemporaries, and historical facts and evidence. Then, there are the different modes and characteristics of people's movements to which Seneca refers, with the 'children, wives, and aged and worn-out parents'; for some, this was a matter of movements without precise destinations; for others to deserted spaces, or spaces occupied by other populations to be conquered 'by force of arms'. But if migration is inherent to humans, what are the direct causes that set them in motion?

Well, people migrate because they have 'lost their own lands', because they are driven out by conflicts, or by natural curses such as plagues and earthquakes. Or because of factors that today would be called Malthusian – 'an excess of population, in order to relieve the pressure at home' – or because they have been 'seduced by the fame of a fertile and over-praised clime'. Lastly, Seneca did not neglect to mention that migrations ensure the renewal and turnover of societies, because 'new names of nations arise, while the former ones die out, or become absorbed by more powerful ones'. Today's scholars are eager to explain the causes of migration with models and algorithms, weighing and measuring the 'pull' and 'push' factors, or the costs and benefits resulting from a change of abode. They, like Seneca, are driven by intellectual curiosity about a phenomenon whose innermost content has not changed so much over the millennia.

I cite these passages from Seneca not as a rhetorical artifice, but as a reminder that today's migratory phenomena unfold through modes and mechanisms that are similar in content, if different in form, to those of two thousand years ago. Historical reflection thus provides essential fuel for reflection on the present. This means comparing people's different motivations for moving, and the forms and modes their migrations take; the existence of selective factors in determining who migrates; the ability of migrants, both individually and in groups, to take advantage of migration; and the mutual benefits of migration, for migrants and for the communities that receive them. Thinking through these aspects allows for a better understanding of migration as a phenomenon even when (as is true of almost all past cases) we lack information that we would today consider essential: how many migrants there are, what their demographic and social characteristics are, where they come from and where they are going. Antiquity provides an opportunity to reflect on a variety of models of migration.

We know very little about the numerical dimensions of the

migration movements that took place over the centuries of ancient history, which we have thus far only briefly glossed over. Moreover, the circumstances, modes and times that characterized human mobility over these centuries varied in the extreme: they ranged from a gradual spread, determined by the natural evolution of communities and peoples in relation to the territory they occupied, to rapid migrations, even over very long distances, of entire populations in search of new settlements. The phenomenon of mobility was thus articulated in a great variety of ways. In these pages we shall attempt to identify some of the most typical (because they have been replicated throughout history) and frequent of them.

As in all eras, there was individual mobility, which is to say, that determined by factors related to the individual, or to his family, or clan, associated with the search for better survival or living conditions. We could call this 'free mobility', usually of short range, and above all typical of relatively homogeneous ethnic contexts. Such a mobility rarely leaves traces, but presumably is related to the urban and commercial development of populations. We may imagine that the flourishing of Greek and Phoenician trading hubs in the Mediterranean – arrival points for the traffic of goods by sea – fostered a particular form of individual migration, as did the stopover stations for overland caravans. Or, furthermore, that matrimonial exchanges, centres of religious devotion and the development of itinerant professions generated opportunities for movement. We can call this type of mobility 'free' because the individual's choice to move is an essential factor in it, if not the only one.

The foundation of the Greek colonies – ἄποικοι – is a model of organized migration, through the twinning of mother cities and the settlement of their citizens in distant lands, for political, economic or Malthusian reasons. The new colonies and cities, in turn, sowed the seeds of further settlements. In the Roman Empire, mobility and migration received a strong impulse from the central political authorities and the army, through the distribution of land to veterans and the formation of new colonies. The Roman *limes* or frontier – that along the Rhine as well as that along the Danube – manned by tens of thousands of soldiers, with their fortifications and satellite and service settlements, constituted another engine of mobility. In addition to their defensive role, they also had another function of catalysing exchanges and mixing with barbarian populations. Beyond the *limes* there were no few cases of forced migration, as Roman militias sent groups and tribes on their way for reasons of defence and control. In the late phase of the empire, the barbarian peoples, whose mobility and competition had been constant, gave rise to forms of collective migrations of entire peoples: those of the Goths, the Huns and the Lombards, known to us also from the writings of contemporary historians such as Ammianus Marcellinus and Paul the Deacon.

1.2. Settlers and founders: *ápoikoi* and *oikistés*

Seneca wrote to Helvia that the 'whole coast of Italy which is washed by the Lower [Tyrrhenian] Sea is a part of what was once "Greater Greece" [*Magna Graecia*]'. Yet the expansion of Greek civilization, from the eighth century BC onward, encompassed the coasts and islands of the eastern Mediterranean, Asia Minor, the Black Sea, the Italian peninsula and its large islands, up to the Mediterranean seaboard of Iberia. This brought various cases of settlement through migration, often in an organized form, whether owing to demographic growth and land scarcity in the motherlands; commercial needs; or political strife and internal conflicts. The emigration process took place in forms moulded by long previous experience. The settlement and founding of colonies by the ἄποικοι took place under the leadership of a chosen prominent personality – an οἰκιστής (*oikist*). There were criteria for the selection of migrants and modes were followed which were supposed to maximize the success of the new colony, which maintained close contacts with the motherland even after its foundation. Many settlements had the character of a trading outpost (ἐμπόριον), whereas others were stable settlements that, in turn, gave rise to further settlements. Thucydides describes the history of settlements in Sicily:

> The first Greeks to colonize Sicily sailed from Chalcis in Euboea, with Thucles as their leader, and founded Naxos. They set up an altar to Apollo Archegetes which still stands outside the city (and delegates to festivals make sacrifice at this altar before they sail from Sicily). In the following year Archias of Corinth, one of the Heracleidae, founded Syracuse, first driving out the Sicels from the island of Ortygia. This, no longer now completely surrounded by water, is the site of the inner city: some time later the outer city was included within the walls and its population grew large.[5]

Thucydides does not tell us whether Naxos was founded in agreement, or in conflict, with the local populations, but the foundation of Syracuse took place violently, and the city prospered 'and its population grew large'. The native peoples could not rest easy and in fact: '[i]n the fifth year after the foundation of Syracuse Thucles and the Chalcidians set out from Naxos, evicted the Sicels by force of arms, and founded first Leontini, then Catana'.[6]

As mentioned, contacts with the motherland were maintained – as confirmed by the following passage about the departure of colonists from Megara, who founded Megara Hyblaea in Sicily, and who after a hundred years – following the *oikist* Pammilus who came from the mother-city of Megara – founded Selinus. The foundation of Megara Hyblaea, it should be noted, took place at the invitation of King Hyblon who, presumably, wanted to profit from its lands:

At about this same time Lamis arrived in Sicily bringing colonists from Megara. He settled a place called Trotilum on the river Pantacyas, but later moved from there to join the Chalcidian community in Leontini for a short while, until they expelled him. He then went on to found Thapsus, where he died. His colonists uprooted themselves from Thapsus and founded the city known as Megara Hyblaea when Hyblon, a Sicel king, in betrayal of his own people made them a gift of the land and escorted them to it. ... Before this removal, and a hundred years after their own foundation, they sent out Pammilus to found Selinus: he had come from their mother-city of Megara to help them establish this new colony.[7]

Groups of colonists, led by their oikists, also arrived from the islands; they came from Crete and Rhodes, or from Cumae, a Greek settlement close to Etruria:

> The foundation of Gela, in the forty-fifth year after Syracuse was founded, was a joint enterprise by Antiphemus from Rhodes and Entimus from Crete, each bringing their own colonists. ... Zancle [Messina] was originally settled by raiders who came there from Cumae, the Chalcidian city in Opicia. Later they were joined by a substantial number of colonists from Chalcis and the rest of Euboea who shared in the distribution of land: the founder-colonists [*oikists*] were Perieres from Cumae, and from Chalcis Crataemenes.[8]

The foundation of the colonies took place in a process that spanned the eighth to sixth centuries BC, following similar patterns. There was a mother-city and a colony; there were close relations and political and trading ties between them, but the colony was independent of the mother; the founder was a notable, the *oikist*, who organized the transfer of the colonists, ostensibly composed of family units. We know nothing about the numerical dimensions of the first groups of settlers, but presumably they amounted to several dozen families, capable of independent survival. We do not know if and how they were recruited, whether this involved selection, who carried out the selection (the *oikist*, perhaps?), how the land was distributed among them, how many colonies survived and how many collapsed, and in what manner. But in the first century BC, Cicero saw a Mediterranean dotted with Greek-founded colonies, in 'Asia, Thrace, Italy, Sicily, and Africa', all 'washed by the waves', 'as if the lands of the barbarians had been bordered round with a Greek sea-coast'.[9]

There are no solid grounds for evaluating the demographic profile of this migratory and settlement process. The new centres that were founded numbered in the hundreds, in a crescendo that reached its peak in the sixth century BC; the size of their population remained modest, as did that of the Greek world in general. We do not know how far the growth of the largest centres – those that exceeded 5,000 inhabitants and even, in the case of Athens and Syracuse, almost 100,000 – owed to natural growth, immigration or the capture of slaves. However, the geographical extension of the city-colonies – from the eastern coast of the Black Sea to the Mediterranean seaboard of Iberia – their number, the intensity of trade and commerce, advances in navigation and other documentary and literary evidence suggest that mobility was very high.

Around 700 BC, there were 23 or 24 colonies in southern Italy and Sicily; it can be imagined that within around fifty years the number of migrants may have exceeded 10,000 (it has been estimated that Megara Hyblaea was founded in 728 BC with two to three hundred colonists).[10] These were small numbers, but with great growth potential. After all, the size of the urban centres from which the settlers arrived over the centuries in question was growing, but still remained of the order of a few thousand inhabitants. Plausible estimates, based on the inhabited area and other objective parameters, assign the Athens of 500 BC some 20,000 inhabitants: 'By 431 BC

Athens probably had 40,000 residents, and its harbor town Piraeus another 25,000. Fifth-century Syracuse was roughly the same size as Athens, and a century later had between 50,000 and 100,000 inhabitants'.[11] These were the two largest metropolises of Greek antiquity. If only for demographic reasons, the migratory flows of this era must have been small in number, but with highly important consequences in the medium and long term.

What is known – not much to tell the truth – about migration in the classical era offers us an interesting model. These were organized migrations, decided upon by the communities of origin that chose the founder and presumably determined criteria for identifying and selecting migrants. The mother communities had to possess good knowledge about the territory that was to be settled, the nature of the available land and the size and attitude of the indigenous populations. The mother communities gathered together (or supplemented) the necessary resources for sea transfer: the ship(s); animals and working tools; seeds and food supplies. It is also to be assumed that the migrants had to be able to cope with the possible hostility of the indigenous populations. It is possible that cases of migration driven by environmental constraints, or famine, or conflict, were not uncommon. But on the whole, this phenomenon appears to have been an activity of expansion and investment that was expected to produce commercial and political returns.

1.3. Augustus's *Res gestae*

Wars, both civil and foreign, I undertook throughout the world, and when victorious I spared all citizens who sued for pardon. The foreign nations which could with safety be pardoned I preferred to save rather than to destroy. The number of Roman citizens who bound themselves to me by military oath was about 500,000. Of these I settled in colonies or sent back into their own towns, after their term of service, something more than 300,000, and to all I assigned lands, or gave money as a reward for military service.[12]

A few years before his death, Augustus wrote a celebratory review of his achievements (*Res Gestae Divi Augusti*), intended to be displayed in the mausoleum which he had built for himself and his family. Two short paragraphs (numbers 3 and 28) out of the 35 that make up the document, offer elements to delineate a type of mobility, and migration, different from that which animated the Greek world. The first of the two is given at the beginning; the round figures quoted by Augustus appear on first sight to be very high, but they belong to an order of magnitude which is considered plausible by the historiography and compatible with many other well-established elements. In the final stages of the civil wars, Rome maintained an army of 60 legions, each of which had a strength of 5,000–6,000 legionaries, providing for a total of over 300,000 not counting auxiliary troops. The legionaries were Roman citizens, subjected to a long term of service (in the time of Augustus it was raised to 16 years) at the end of which they were compensated with sums of money or land. The interpretation of census figures is open to debate. But one indicator used by many scholars is that the population of Italy in Augustus's time consisted of 5 or 6 million inhabitants (including 1 to 1.5 million slaves), with about a quarter of its citizens able to take up arms.[13] It thus sounds entirely possible that half a million Roman citizens had served with Augustus. The reported figure of just over 300,000 covers the veterans who had returned home (many had presumably died or gone missing, or deserted, before the end of their service) or whom the emperor had 'sent to the colonies', that is, to found new settlements in various parts of Rome's dominions. Indeed

> I settled colonies of soldiers in Africa, Sicily, Macedonia, both Spains, Achaea, Asia, Syria, Gallia Narbonensis, Pisidia. Moreover, Italy has twenty-eight colonies founded under my auspices which have grown to be famous and populous during my lifetime.[14]

How many people went to form the new colonies (the Italian ones included Aosta, Turin and Trieste) is unknown. But this was certainly a matter of many tens of thousands of people, who together with their families and slaves must have amounted to a few hundred thousand. We are not interested here in fixing a number, even a very approximate one, but rather in pointing out how the army was one of the driving forces – perhaps the main one – of the migratory processes in the Roman world, particularly in the final phase of the republic. Almost a century ago, Rostovtzeff commented as follows on the periodic land redistributions that occurred during the civil wars:

> According to careful calculations, not less than half a million men received holdings in Italy during the last fifty years of that troubled period. After the great changes of the 'Social' war, these redistributions were perhaps the most potent factor in the Romanisation and Latinisation of Italy.[15]

During the second triumvirate of 43–33 BC, another great colonization plan was made for the veterans,

> such that the new military colonisation under the triumvirate came to affect slices of land that had been left to the local populations . . . in order to make room for the veterans, who had to be gratified in every way, since the success of future war campaigns and the very possibility of new enrolments depended on the satisfaction of their needs.[16]

There were 170,000 veterans to be accommodated, and at the end of the civil wars there would need to be further allocations of land for the victors and perhaps even for the veterans of the defeated side, if pacification was indeed sought. But also to be borne in mind was 'an equally large number of persons expelled from the land on which they lived and worked. The expulsions and expropriations were legal and licit . . . but the social consequences of the operation that was being carried out could be severe and cruel.' It is not that other forms of mobility were absent: in particular there were forms of forced migration, as in the cases of slaves, prisoners of war and those condemned to work in the mines. But the recruitment, deployment and resettlement of soldiers (often far from their homes) was perhaps the most prevalent form of mobility. According to some authors, massive dislocations of populations from one part of the Roman world to another, according to political or economic expediency, were not uncommon. This had already happened in previous centuries in the great autocratic empires such as Egypt or Persia.[17]

The Roman case differs significantly from the Greek one, although the results were similar in both instances, despite the very different contexts. Over the centuries, both the Greek cities and Rome promoted and created a web of well-established and structured settlements that extended and strengthened commercial and cultural exchanges with areas hitherto beyond their direct control; this produced ethnic mixing and enriched the network of cities across the Mediterranean. In both cases, migrations were organized and guided, either by the mother-city or by the state, which bore the cost, at least in the initial phase. In the case of Rome, however, this movement was guided and organized by a centralized state with precise aims and directions. In the case of Greek cities, the foundation of sister colonies also had competitive features with neighbouring or rival cities. In the case of Rome there was a clear selection of founders, almost all of whom were veterans, men steeped in conflict and discipline. In the Greek case it is to be assumed that the first settlers were – while certainly not a representative sample of the population – made up of elements from a variety of social and professional categories. The Greek colonists who founded settlements had somehow to coexist, or contend and compete, with the local populations, while the Roman ones settled in secure lands under the control of the central state.

It was along the frontiers, the *limes* again, of the empire that Rome's central political authorities acted as a driver of other migrations, by means of army functions and activities. A contemporary historian has aptly summarized the migration issue in the Roman Empire: this was a people with sharp internal inequalities, but one that had a strong basis in

> a stable administration and an integrated economy; on the outside peoples forced to survive with insufficient resources, threatened by famine and war, and increasingly asking to enter; a militarised border to filter out refugees and migrants; and government authorities that had to decide on a case-by-case basis how to behave towards these emergencies, with a range of options from forced removal, to mass reception, from setting entry quotas to offering humanitarian aid and jobs.[18]

The Roman world, in all its complexity, offers many interesting insights into human mobility. A world that over the course of many centuries expanded to encompass the entire Mediterranean, reaching as far north as Britannia, gave rise to the most varied forms of mobility and migration. The state, as we have already said, played an important role in this, and we should note two aspects in particular. The first concerns the management and operation of the borders, which were largely militarized, in the dual function of barrier and filter. The army, positioned in order to guard the empire's borders, thus played an important role in the Roman migration system, both on account of the defensive and contentious relations with 'barbarian' populations on the other side of that border, and for its function of catalysing urbanization and mixing with the indigenous populations – not always allies and friends – in Rome's territories. The second aspect relates to the upheavals among the external populations – those not allied or federated with the empire. They were obliged to mount forced migrations and mass departures, as dictated by Rome's need to defend itself from invasions, to tame more aggressive peoples and to establish buffer territories.

The *limes*, established with the main purpose of blocking unwanted immigration and invasions by peoples outside Rome's jurisdiction, were at the same time also a factor for mobility and exchange. The border set out along the Rhine was 1,300 kilometres long; the one along the Danube extended more than twice as far. The *limes* were dotted with fortresses, entrenched camps, troop stations and garrisons controlled by the army. In the first century AD, some eight legions controlled the 1,300 kilometres of the *limes* along the Rhine, with a presumed total of 40–50,000 men, plus auxiliary troops; the Danubian *limes*, being much longer, was defended by twelve legions (60–70,000 men) in Trajan's time, not counting the numerous auxiliary forces. Around the fortresses and fortified camps arose civil agglomerations (*canabae*), which attracted a varied human canvas consisting of merchants, artisans, tavern-keepers, prostitutes, jugglers and slaves. These agglomerations turned into permanent settlements, even of an urban nature, and were meeting places between garrison men and natives. But the Germanic peoples settled on the opposite banks – east of the Rhine and north of the Danube – often had friendly relations with the Roman military and their retinue; the borders were often infiltrated by groups of barbarians eager for better living conditions:

> The community next adjoining, is that of the Hermondurians; (that I may now follow the course of the Danube, as a little

before I did that of the Rhine) a people this, faithful to the Romans. So that to them alone of all the Germans, commerce is permitted; not barely upon the bank of the Rhine, but more extensively, and even in that glorious colony in the province of Rhoetia. They travel everywhere at their own discretion and without a guard; and when to other nations, we show no more than our arms and encampments, to this people we throw open our houses and dwellings, as to men who have no longing to possess them.[19]

It can be presumed that these complex relations between the two banks gave rise to more or less prolific unions between native women and Roman soldiers, who were forbidden to marry or to bring their wives with them. Given the length of the borders, the multiplicity of barbarian populations and ethnic groups, and the variety of situations, whether of peace or conflict, relations between Romans and indigenous populations were extremely varied, and cannot be reduced to a single model. The Hermondurians were friends of the Romans, while the Sugambrians, who had been in conflict for decades, were defeated by Tiberius who 'brought forty thousand prisoners of war over into Gaul and assigned them homes near the bank of the Rhine'.[20] To the east 'Sextus Aelius Cato settled 50,000 "Getae" who were probably Dacians, south of the Danube, in what was later to become the province of Mesia'.[21] Leaving aside the figures – which are notoriously unreliable and prone to exaggeration – the populations displaced and transplanted from one place to another were of considerable size. These were forced migrations dictated by the state, out of strategic and political considerations, in the interest of the state itself. As for the mobility caused by the management of the *limes*, this involved not only conscripts, but also volunteers who found personal gain in enrolling, and who moved of their own free choice.

1.4. Peoples on the march

> Alboin, being about to set out for Italy with the Langobards, asked aid from his old friends, the Saxons, that he might enter and take possession of so spacious a land with a larger number of followers. The Saxons came to him, more than 20,000 men, together with their wives and children, to proceed with him to Italy according to his desire. Hearing these things, Chlothar and Sigisbert, kings of the Franks, put the Suavi and other nations into the places from which these Saxons had come.[22]

Beginning in the third century, the pressures of the Germanic peoples on the empire's frontiers intensified. History allows us to see a kaleidoscope of peoples and ethnic groups, almost always of uncertain origins, their history shrouded in fog, with experiences of conflict and intermingling, of largely nomadic character, in vast and sparsely populated territories. It may be presumed that over time their numbers increased and that many were making a slow transition from nomadism to a settled existence. On the other side of the *limes* stood a large and populous, well-organized empire, with a much higher standard of living, knowledge and technology. The growth of the empire, its recurring crises and the multiple barbarian pressures on the frontiers required the adoption of a flexible policy, capable of admitting settlements of peoples within the *limes* when it suited, of integrating foreigners into the army to bolster its ranks when necessary, of bargaining with neighbouring tribes and of suppressing raids and invasions.[23] Up until that point the empire had shown itself able to handle the pressure from outside peoples, particularly the Alamanni and Franks, along the Rhenish *limes*. Alessandro Barbero cites the so-called *Panegyric in Honour of Constantius* (Constantius Chlorus, father of Constantine) written after his victories over the Franks who had crossed the Rhine and invaded the delta lands in Gaul with raids, before being driven back across the river or taken prisoner and deported:

> In all the porticoes of our cities sit captive bands of barbarians, the men quaking, their savagery utterly confounded, old women and wives contemplating the listlessness of their sons and husbands, youths and girls fettered together whispering soothing endearments, and all these parceled out to the inhabitants of your provinces for service, until they might be led out to the desolate lands assigned to be cultivated by them ... And so it is for me now that the Chamavian and the Frivian plows, and that vagabond, that pillager, toils at the cultivation of the neglected countryside and frequents my markets with beasts for sale, and the barbarian farmer lowers the price of food, and they come to sell their livestock in my markets, and it is a barbarian farmer who pays the tax. Furthermore, if he is summoned to the levy, he comes running and is crushed by discipline; he submits to the lash and congratulates himself upon his servitude by calling it.[24]

It can be inferred from the *Panegyric* that the empire's ability to contain the pressures of barbarian peoples remained intact, allowing it to take advantage when barbarians could be settled in depopulated areas, either by force or as a result of agreements. From the *Panegyric* it can also be deduced that deportation concerned not only men or warriors but entire peoples – old men, women and children included. Indeed, it is quite plausible that the mobility that characterized the lands inhabited by barbarians, to the north and east of the empire, usually involved whole peoples.

To the east, the Danubian *limes* separated the empire from various barbarian peoples, among whom the Goths especially stood out, whose pressures on the frontier and raids across it had been effectively contained during the reign of Constantine the Great (306–337). The Goths were in the process of Christianization, and were in frequent contact with the Romans. The situation changed rapidly with the arrival of the Hunnish people, coming from the eastern steppes, who overwhelmed many ethnic groups and tribes in their path before approaching the Danube. A century later the Byzantine chronicler Zosimus recounted the Huns' approach in 373 as follows:

> I have met with a tradition, which relates that the Cimmerian Bosphorus was rendered firm land by mud brought down the [river] Tanais, by which [the Huns] were originally afforded a land-passage from Asia into Europe. However this might be, they, with their wives, children, horses, and carriages, invaded the Scythians [Goths] who resided on the Ister [Danube] ... [And] though they were not capable of fighting on foot... by the rapidity with which they wheeled about their horses, by the suddenness of their excursions and retreat, shooting as they rode, they occasioned great slaughter among the Scythians. In this they were so incessant, that the surviving Scythians were compelled to leave their habitations to these Huns and cross the Ister.[25]

These were whole peoples migrating, with women, children, animals and baggage, though they were also ready to settle along the way. Presumably even the Goths[26] who crossed the Danube in dramatic circumstances were a 'people' and not just warriors. A contemporary historian, Ammianus Marcellinus, a military man and a champion of Rome's civilizing mission, reported that the Goths, pressured by the Huns, sought the emperor Valens's permission to cross the river and settle peacefully in Roman lands: 'and sent ambassadors to Valens, asked with humble prayer to be received, promising to live quietly, and to also provide to him with relief when circumstances required it.' Valens granted passage, thinking that the Goths would reinvigorate the army and settle in the unproductive lands of Thrace. He sent relief and chariots to transport the savage horde, and diligent care was taken that no future destroyer of the Roman state should be left behind, even if he were smitten by a fatal disease. Accordingly, having by the emperor's permission obtained the privilege of crossing the Danube and settling in parts of Thrace, they were ferried over for some nights and days embarked by companies in boats, on rafts, and in hollowed tree-trunks; and because the river is by far the most dangerous of all and was then swollen by frequent rains, some who, because of the great crowd, struggled against the force of the waves and tried to swim were drowned; and they were a good many. With such stormy eagerness on the part of insistent men was the ruin of the Roman world brought in. This at any rate is neither obscure nor uncertain, that the ill-omened officials who ferried the barbarian hordes often tried to reckon their number, but gave up their vain attempt; as the most distinguished of poets says:

Who wishes to know this would wish to know
How many grains of sand on Libyan plain
By Zephyrus are swept[27]

What was supposed to be a peaceful migration quickly took on a quite different hue; the relief promised by the emperor did not arrive or was intercepted by corrupt soldiers; the emperor's benevolence turned into a strict ban on new Goth migrants, who crossed the river anyway. The situation changed radically, through the raids mounted by the Goths and the armed clashes, producing a conflict that got out of control. This reached its tragic conclusion in the battle between the Goths and the Roman army at Adrianople (378 CE), which ended with the defeat of the latter and the killing of Valens.

The chronicles are confused, inaccurate and, as always, unreliable in terms of numbers. But a contemporary scholar estimates that the Gothic warriors (Ostrogoth and Visigoth) who defeated Valens at Adrianople numbered many thousands, and that Gothic populations capable of fielding 20,000 warriors had crossed the Danube – a figure that could indicate a substantial population of between 50,000 and 100,000.[28] Regardless of the number and role of the Goths in Rome's territory in the following century, this was the first mass barbarian invasion across the *limes* of the empire, which ended with the fall of the Western Roman Empire one hundred years later. But the Gothic domination would also collapse, more than two centuries later, due to the arrival from the north of another migrant people, the Lombards, whose history is narrated to us by Paul the Deacon:

> the Langobards, having left Pannonia, hastened to take possession of Italy with their wives and children and all their goods ... when king Alboin with his whole army and a multitude of people of all kinds had come to the limits of Italy, he ascended a mountain which stands forth in those places, and from there as far as he could see, he gazed upon a portion of Italy.[29]

II
In the Hands of the State

2.1. Forced migration

Moving and shifting across territory are human prerogatives. Inscribed in the biology and sociality of individuals, these practices are functional to the pursuit of well-being and survival. Migration, when it is voluntary, represents a concrete exercise of these prerogatives. But its voluntariness is never absolute, and stands as the positive pole of a spectrum that has its opposite, negative pole in forced migration, including cases of expulsion, displacement or deportation. These take different forms, depending on the context and the historical era; here, we are referring to settled populations that are uprooted from their homes and forcibly transplanted elsewhere.

The examples presented here do not include the most imposing case of forced movement in the modern history of the Western world: namely, the deportation of slaves from Africa, which has profoundly affected the ethnic and human geography of the American continent. Its exclusion may be justified, in a formal sense, by the fact that these pages are dedicated to Europe and the Americas, and not to the other continents. But that is a very weak justification. Indeed, the traffic in slaves has had enormously important effects on the Americas; suffice to mention that some two hundred million people, out of their one billion inhabitants, today call themselves either 'Blacks', north of the Rio Grande, or 'Afrodescendientes' south of it. However, the real reason for excluding this example is that the slave trade is a historical fact of such scope, duration and dimension, and of such tremendous cruelty, that it requires an exclusive treatment that finds no fitting space here. Two centuries ago, Alexander von Humboldt wrote that 'Slavery is possibly the greatest evil ever to have afflicted humanity', having observed many aspects of it during his five-year exploration of South America.[1]

A few quick reminders of some aspects of trafficking. Careful studies of archival material have concluded that between 1500 and the latter part of the nineteenth century – when trafficking, which had become illegal earlier that century, ran out of illegal arrivals –between ten and eleven million Africans were disembarked in American ports. But these were only the ones who survived the terrible conditions of travel on the slave ships, and in fact the number of slaves embarked in Africa was considerably higher.[2] About two-thirds of the slaves were young adult men (*piezas de India*),[3] and the remaining one-third women and children. The slaves came from the west coast of Africa, from Senegal to Angola, and a small minority also from the east. About half were taken to the Caribbean islands owned by the Spanish, French, British and Dutch; a good third to Brazil; and the remainder to the southern part of the present-day United States and the South American colonies, later ex-colonies, of Spain. About two-thirds of the slaves were brought to America between 1750 and 1850.

The transatlantic slave trade was not undertaken for the purposes of populating the American continent, but rather to provide a cheap, mostly male, workforce for production and services, which could supplement or replace a native workforce that was too scarce and in sharp decline after contact with Europeans. It is almost as if this were a raw material, forcibly extracted from the African continent to feed the New World, according to the needs of the market. The enslaved population had a very high mortality rate and – given the hindrances placed on family formation and reproduction – a very low birth rate; it had to be fed by new arrivals to prevent its numbers from declining. On the large sugar-cane plantations in the Caribbean and Brazil, it was widely believed that it was cheaper to buy slaves on the market to fill gaps, rather than to invest in the welfare of the enslaved people and sustain their reproduction.

The three cases of forced migration narrated here are very different. The first, concerning the Inca Empire, is the least traumatic: it is about forced transplants of entire communities made to consolidate the new territories acquired in the process of imperial expansion in the fifteenth and sixteenth centuries, and to dilute and integrate the populations inhabiting them. These displacements were not undertaken for punitive purposes: indeed, the communities forced to migrate retained many of their prerogatives and also acquired new ones. But they were still forced displacements. The Spanish also subjected Peruvian populations to a radical process of displacement in order to alter settlement patterns in the colony and thus achieve a better social control of the population. The second case concerns the Ottoman Empire, in its twilight years, before, during and after its end. This twilight was characterized by the internal displacement of minorities and population exchanges with neighbouring states. In these tumultuous years, forced migrations resulted in the genocide of the Armenian minority, actions to exterminate the Greek minority, and the forcible exchanges of large communities: Greeks from Anatolia to Greece, and Turks from Greece to their homeland. Religious factors weighed heavily in the Ottoman case, and through Europe's history they would also prompt other forced migrations: conspicuous examples include the expulsion of the Jews and then the *Moriscos* from Spain and the exodus of the Huguenots from France. In the third case, concerning the Soviet Union during the Second World War, these were 'internal' forced migrations, or rather outright deportations, of minorities suspected of consorting with the enemy or otherwise deemed politically unreliable. The ethnic groups of Koreans, Germans, Finns, Tatars, Chechens, Ingush, and many others, were relocated to distant and peripheral areas of this country's boundless territories. More generally, the two world wars, with the redrawing of the map of Europe, population exchanges aimed at ethnic homogeneity, forced migrations and waves of refugees generated massive population displacements which escaped individual willingness or choice.

In modern times, population removals are relatively rare; more frequent are the expulsions and flight caused by conflicts and violence involving millions of people, officially considered 'refugees' by international institutions. In 2021, the number of refugees forced to leave their homelands, according to the United Nations' special agency (UNHCR), amounted to 27 million, plus 48 million people forcibly displaced within their own countries. This is a kind of pathology of mobility, which increased conflict has turned into a mass phenomenon. If we consider that according to UN assessments, the total inventory of migrants in the world (people living in a country other than their land of birth) counts not far from three hundred million people, we can say that for every ten migrants there is one who is forcibly displaced outside their own country. In the past, the partition of the British Raj (Anglo-Indian Empire) between India and Pakistan in 1947 resulted in a number of refugees variously estimated at between ten and twenty million; in 1983, the expulsions of Ghanaians and other irregular West African migrants from Nigeria affected more than two million people; the wars in Yugoslavia in the decade between 1991 and 2001 generated more than four million refugees and internally displaced persons. In the last decade, the expulsion and deportation of a million Rohingya from Myanmar to Bangladesh is still an open wound, as is that generated by the countless refugees from the civil war in Syria and Russia's attack on Ukraine.

History is replete with examples of forced migration. While the great variety of situations precludes any real counting of the numbers, there is no doubt that the passing of time has been unable to alleviate the pathologies that cause it.

2.2. Peru: up and down the Andes

> The Inca took Indians from Nanasca to transplant them to the banks of the river Apurímac, because that river, from the royal road that passes from Cuzco to Rimac, flows through such a hot region that the Indians of the sierra, from a cold or temperate climate, cannot live in such heat, but fall ill and die. So, as already mentioned, the Incas stipulated that, when they transplanted Indians from one province to another in this way, what they called *mítmac*, they would always draw a comparison between the regions, making sure that the climate there was the same, so that the difference in conditions was not harmful, such that moving them from a hot region to a cold one, and vice versa would not cause them to die. For this reason, it was forbidden to bring Indians from the sierras to the plains because, with all certainty, they would perish within a few days. (Garcilaso de la Vega)[4]

In ancient Peru, despite the ruggedness of the terrain, mobility was not a problem: it is said that the Inca in Cuzco could every day enjoy fresh fish caught in the ocean more than three thousand metres below. This is probably a myth, but the ingenuity of the Inca people and the ever-expanding size of the empire, as well as the need to govern it, had endowed the country with a road network second only to that of the Roman Empire, from sea level to the highlands over 4,000 metres above sea level. To consolidate such an extensive empire, formed in a relatively short period (essentially in the fifteenth century), the Incas often resorted to the practice of forced migration of the populations there. Pedro Cieza de León, who had headed to the Indies at a very young age, travelled the whole of Peru as a soldier for twenty years and was then officially appointed *cronista de Indias*.[5] A broad-minded chronicler whom the historiography considers reliable, he wrote in the middle of the century:

> They call *mitimaes* those who are transferred from one land to another; and the first manner of dealing with *mitimaes* established by the Incas was that, after a province had been conquered or reconquered, in order to keep it safe, and so that in a short time the natives might learn how they should serve and behave, something which their vassals had long understood and known, and so that they might be peaceful and quiet, and not often rebel – and if they did, there were those who prevented them from doing so – they transferred to those provinces as many people as it seemed opportune to transfer; they commanded these latter to populate another land, with the climate and conditions of the one which they had left, if cold, cold, if warm, warm, in which they gave them as many lands, fields and dwellings as those which they left; and from the lands and provinces which they had long considered peaceful and friendly, and of whose willingness to serve they had experience, they commanded as many or more to go out and settle in the newly conquered lands and among the Indians, in such a way that they depended on these latter for the things we have said above, and kept them in good order and cleanliness, so that through the departure of some and the arrival of others, everything would be safe with the governors and delegates who settled.[6]

This internal migration was imposed and planned from above for various reasons, such as the consolidation of acquired territories, the defence of borders, the integration of new subjects into the system and economic suitability. In many cases, this meant long-range migration, in an empire that at the time of the conquest spanned 4,000 kilometres, from southern Colombia to mid-Chile.

Garcilaso de la Vega, the son of an Inca princess and a conquistador, who had left Peru for Spain at the age of twenty-one (more than half a century after Cieza de León), wrote:

> As they proceeded in their conquests, the Incas came across provinces that were fertile and rich, but very sparsely populated, and for this reason poorly cultivated; and in such provinces, in order not to lose their fruits, they transferred Indians from others, of the same type of climate, whether cold or hot, so that they would not have to suffer the difference in conditions. At other times they would do this when their numbers multiplied so much that they could no longer stay in their provinces; they would then seek out others suitable for them; they would bring half of the people from the said province, more or less, according to convenience. They also removed Indians from poor and barren provinces, in order to populate fertile and abundant lands, and this was to the advantage both of those who left and those who remained, so that, belonging to the same stock, they might help each other in their harvests, as was necessary throughout the Collao, a province which measures more than a hundred and twenty leagues in breadth, and includes many other provinces of different nations and, because it is a very cold land, neither maize nor *uchu*, which the Spaniards call *pimiento*, grows there, while there is a great importance of other grains and legumes, which are not found in the hot lands, such as th[ose] which they call *papa y quinoa*, and the livestock that graze there are very abundant.[7]

The system was surely widespread in the great empire, and the Spaniards noted its scope and complex economic, cultural and legal implications. There were also incentives for the *mitimaes*, including:

- the distribution and allocation of land in the regions of immigration;
- allotments of 'provisions, as many as were necessary, until they obtained the fruits of their crops';
- exemption from tribute 'before they were provided with women, and coca, and resources, so that they might more willingly work among their peoples';
- the maintenance of contacts and rights acquired in the lands of origin.

The policy of transplantation, or dislocation, or forced migration of the Incas was very complex, motivated by concerns about defence, about controlling ethnic groups considered rebellious or untrustworthy, and others of an economic or administrative nature. There are various and widespread pieces of evidence of this, especially in the Bolivian highlands.

The system of more or less compulsory migration imposed by the Incas overlapped with a highly articulated underlying regime, which has been defined as a 'system of vertical control that extended to a maximum number of ecological levels' by the various ethnic groups.[8] This is the case, for example, of the Lupaca people, a group settled in the Titicaca basin (Chucuito district) that numbered around 100,000 inhabitants at the time of the Spaniards' arrival. The economic and social organization of the Lupaca, at an altitude of almost 4,000 metres, was based on a few staple crops (quinoa, potato) and involved control of trade and stockpiling. This included the control of Lupaca groups settled in the valleys sloping down towards the coast, some hundreds of kilometres away, where cotton and maize were cultivated, or even lower down, where pimiento and fruit were grown, complementing the highland crops. In other cases, the original groups were much less numerous and controlled small settlements a few days' walk away, either at higher altitudes – used for grazing – or at lower altitudes, for coca cultivation, cotton, salt production and more. In any case, the geographical structure of the country, inhabited from the seaboard up to an altitude of 4,000 metres, benefited from the control of those resources which could not be obtained in the centre of the settlement. The inhabitants of the secondary districts retained their rights and dwellings in their population centres of origin. This integrated system certainly involved considerable mobility, and especially a 'vertical' mobility, between the low altitudes and the mountains, between the coast and the highlands. There was thus a vertical human and economic connection, which implied migration and mobility. This meant an orderly, guided mobility, imposed by the community and the state, which prevailed over individual decisions.

When the empire fell into the hands of the Spaniards, there was another very extensive example of forced displacement. It was a process of 'reduction', i.e. dislocation and resettlement of the population, which, having previously been dispersed or aggregated in small centres, was redistributed into larger groups, into new planned and ordered villages: with a grid pattern of streets, a central square, a church and public buildings. This was a process that affected all of Hispanic America during the sixteenth century, and it had three aims: a political one, to better control a population traditionally accustomed to living scattered over enormous and often impervious expanses of territory; a religious one, to ensure their conversion and indoctrination; and an economic one, relating to the identification of taxpayers and the collection of tributes. In Peru, this policy was implemented rigorously and quickly by Viceroy Toledo,[9] who (in 1572–3) instructed his officials on how to conduct this gigantic operation: it was necessary to identify suitable locations, to concentrate the population in as few villages as possible, and to draw the plan of each village, with a grid pattern of roads and blocks. There were also rules for building dwellings, and a ban on dispossessing the Indians of their fields in their original home areas; it was also necessary to build the villages far from the traditional places of worship; to transfer the Indians to the new villages; and raze their traditional dwellings to the ground. During five years of visits and 'peregrinations', Toledo worked to personally supervise and coordinate this operation, and in his *Memoria* for Philip II he wrote

> It was not possible to indoctrinate these Indians or to make them live in an orderly political way without dragging them out of their hiding places. In order to do this, we had to move them to villages and public places and lay out a gridwork of streets as in the Spanish towns, making sure that the doors [of the houses] opened on to the streets so that they could be seen and visited by the authorities and the priests. In carrying out these reductions, we always had as our objective to locate the villages in the best places of the districts and ones that had a climate similar to that from which the Indios came. Moreover, the new villages should have a sufficient number of tributaries to support one or two priests for the work of indoctrination.[10]

The relocation of the population was a colossal operation, involving one and a half million people – a figure that is backed up by specific figures for a considerable number of districts.[11]

The habit of Andean populations of moving around was fully exploited by the Spaniards, with the discovery in 1545 of the extremely rich silver mines at Potosí, which soon became the main source of precious-metal production across the Americas. At the time of discovery, Potosí, standing at an altitude of 4,000 metres, was but one among many insignificant villages. But it was dominated by a 600-metre-high mountain (Cerro de Potosí), which turned out to be extremely rich in veins of silver ore. It was soon perforated by dozens of tunnels. The year after the discovery, there were 300 Spaniards at the foot of the mountain; already by 1547, the year the city was founded, there were 14,000 of them; in 1611, with 160,000 inhabitants, Potosí was one of the most populous, and richest, cities in the western hemisphere, surpassed in the Old World only by London and Paris. But the mining and processing of ore required a large workforce, which was absent in this semi-deserted district. An ancient institution, the *mita*, allowed the recruitment of Indians for *corvée* service in exchange for payment; although personal service had been abolished by Spanish law, it was permissible in the case of works of public utility. The mining of silver was, without a doubt, of public utility, because it replenished the coffers of the colony and the crown, and this principle thus became the keystone of the system designed by Viceroy Toledo, which would remain in force for two and a half centuries, until its final abolition by Simón Bolívar in 1825. The system devised by Toledo provided for the annual recruitment of around 14,000 adult Indians between the ages of fifteen and fifty, equal to about 14 per cent of the tributaries, drawn from a strip of the highlands some 1,400 kilometres long by 400 wide, involving 125 communities. The sending of the *mitayos* (the Indians obliged to serve) took place annually, and the duration of work in the mine was one year. This implied – in theory, and on average – that each of these communities' adult males was obliged to serve for one year out of every seven:

> The people who gathered in this town [Potosí], added to its residents, numbered some 13,340 Indians; but to reach this number more than 40,000 people had to leave from their villages, with their wives and children. The roads were so crowded that it seemed as if the whole kingdom had set in motion.[12]

The distances that had to be travelled were considerable, ranging from a minimum of a few dozen to a maximum of 180 leagues (almost a thousand kilometres) for those coming from the Cuzco district. The Indians travelled with their families and household goods, accompanied by llamas for transport, and congregated at the orders of their *curacas* (village chiefs) at gathering places, before setting out on their way together. This biblical throng of men, women, children and beasts had to make journeys to the most remote places, sometimes even lasting over a month, at a rate of five or six leagues a day (a small fee was allotted for each day's journey). These steps were then retraced in the opposite direction after the year of *mita* service.

The Peruvian case shows the importance of government intervention – be it indigenous or colonial – to human mobility in a great empire. Here, we have spoken of three very different modes of mobility: the frequent but non-systematic one, carried out by the Inca with the *mitimaes*, which consisted of transplanting entire populations; the forced and radical redistribution of the population, made on a one-off basis by the government of the colony; and the perpetual movement of people, working through a sophisticated mechanism, set in motion by Toledo. These modes of mobility each found a certain degree of acceptance among the populations concerned and did not generate conflict, anarchy or violent opposition. These forms of migration were not selective – except, in the first case, in terms of picking the peoples to be transplanted. There was no selective choice of individuals, and these movements instead involved entire communities and whole families. Mostly political motivations guided the transplantation of *mitimaes* and the redistribution of settlements; in the case of the *mita* of Potosí, the motives were mainly economic. We know little about the other forms of mobility – spontaneous ones, or ones in any case not controlled by the authorities – which resulted from the large population of vagrants, fugitives, stragglers and tax evaders, added to the physiological mobility linked to trade, transport and certain particular activities.

2.3. The end of an empire

> Sir! We beseech you in the name of God and human brotherhood – we, the population of seven Armenian villages, in all about five thousand souls, who have taken refuge on that mountain plateau of Musa Dagh, known as the Damlayik, and three leagues north-west of Suedia above the coastline. We have taken refuge here from barbarous Turkish persecutions. We have taken up arms to preserve the honour of our women.
>
> Sir! You no doubt have heard of the Young Turkish policy which seeks to annihilate our people. Under the false appearance of a migration-law, on the lying pretext of some non-existent movement for revolution, they are turning us out of our houses, robbing us of our farms, orchards, vineyards, and all our movable and immovable goods and chattels.[13]

On 8 October 1912, the First Balkan War broke out; it ended a few months later with the defeat of the Ottoman Empire by the coalition formed by Bulgaria, Greece, Montenegro and Serbia, and the almost total loss of its European territories. Ten years later, on 1 November 1922, the sultanate was abolished, forcing the last sultan, Mehmet VI, into exile. The Turkish republic, heir to the Ottoman Empire, which at its height stretched from the Danube to the Maghreb, was restricted to Anatolia and a small European appendage. In this decade of searing defeats, the losses of the war among both civilians and soldiers counted on the order of four million, and there were more than a million victims of indiscriminate massacres against minorities, which the historiography near-unanimously defines as genocides. More than two million others were the victims of forced – or in any case, politically decreed – migration movements, and countless numbers were displaced as a consequence of war and political events. All this in a country that, at the time of the proclamation of the republic, had about 14 million inhabitants, and that had stripped itself of cosmopolitan minorities in pursuit of a greater cultural, linguistic and religious homogeneity. Such homogeneity was achieved by massacring, deporting and expelling those minorities, minorities who had been settled in that territory for centuries or even millennia.

Much is known about the historical events of this decade, though they are the subject of debate and controversy and remain open to further exploration. Less is known about economic and social developments, and even less about demographic ones, due to the imperfection and scarcity of sources. Our focus is the nature of these migration movements (which, it should be repeated, were mostly forced) and the motivations behind them. They were certainly of a cultural and political nature: after all, for them to be translated into action, they presupposed the existence of an organized and strong state (domestically, that is: externally, the country suffered searing defeats). This state also relied on a robust ideology that would be able to mobilize its subjects and citizens. The roots of the events that marked this decade, in terms of migration – especially concerning the Armenian and Greek communities, by far the most numerous – lie in the gradual crumbling of the Ottoman Empire.

As early as 1878, Romania, Bulgaria, Serbia and Montenegro had achieved independence; Bosnia-Herzegovina had been occupied by Austria; this latter exercised a strong and growing influence on the Balkan countries, as did Russia on Bulgaria. The First Balkan War had taken Albania, Rumelia and Macedonia from Turkey, along with almost all of Thrace and the island of Crete. Dissatisfaction and opposition to the empire's repressive policies were growing, and not only among the more mature and open-minded classes: in 1907 the Committee for Union and Progress, which included the Young Turk movement, was founded, and in 1908 it took power. The most radical nationalist component prevailed within this Committee, and in 1913 it hardened into a dictatorship, strengthened the following year by its entry into the war on the side of Germany and Austria-Hungary. As is well known, the war ended disastrously, with three million military and civilian casualties; Allied forces occupied Istanbul, the Greeks occupied Smyrna and its region, Syria went to the British and, with the Treaty of Sèvres in 1920, the empire was effectively dismembered. The new Turkish National Movement, led by Mustafa Kemal (later Atatürk), did not accept the terms of the defeat and the army went into action against the Greek army. After various developments, and reciprocal massacres among the Greek and Turkish civilian populations, the Greek army was defeated, sealing what the Greeks called the 'catastrophe of Asia Minor'. With the Treaty of Lausanne (1923), Turkey was reduced to the Anatolian region, plus its European appendix. Along the lines of Wilsonian principles, it was decided to resolve the centuries-old Greek-Turkish question with a gigantic population exchange between the two belligerents; the Armenian question had already been closed with the destruction of that people.

The empire – in the territories that make up today's Turkey – was rich in ethnic and religious minorities. It has been estimated that in 1914 the population was 18.8 million, consisting of a Muslim majority (80.7 per cent) and sizeable Greek (9.3 per cent), Armenian (8.5 per cent) and other minorities (1.5 per cent).[14] Some of these minorities – in particular Christians (Greek Orthodox and Armenian) and Jews – constituted communities recognized by the state and had enjoyed administrative autonomy for centuries; however, their members were considered a kind of separate class, subordinate to Muslim society. The social growth of these national and religious groups at the turn of the century was viewed with mistrust and resentment, not least because of their proximity to foreign powers, whom, it was feared, were ready to carve up the spoils of the declining empire:

> It was in the course of this drive toward nationalism ... that the idea of a territorially identifiable Turkish homeland, of a

homeland capable of constituting at the same time a mobilising myth, a shared ideal and a hope for the future in which to invest the fears and disappointments of the present, entered everyday public discourse.[15]

This homeland was identified, territorially, with Anatolia, which was now pressured by the Muslim immigration of refugees and displaced persons from the European regions lost by the empire. The entry into the war on the side of the Central Powers, the disastrous opening battles against Russia in the Caucasus mountains, and the widespread sense of danger, provided the ruling triumvirate with its pretext to begin the 'removal' of the Armenian community, which was by now conceived as an enemy. With a series of measures, from the end of 1914, the state expelled Armenians from the army, arrested political and civilian leaders in Istanbul, confiscated their property and assets and authorized the deportation of the population from the eastern provinces where Armenians were concentrated:[16]

> Guarded by Ottoman soldiers and gendarmes, they were attacked and slaughtered by the *çetes* (gangs of irregular fighters) of the *Teşkilât-ı Mahsusa* (Special Organization), and by Kurds, Turks, and Circassians. Driven to exhaustion, starvation, and suicide, hundreds of thousands would perish; others would be forced to emigrate or convert to Islam to save their lives. Men died in greater numbers; many women and children were taken into the families of the local Muslims. Tens of thousands of orphans found some refuge in the protection of foreign missionaries. It is conservatively estimated that between 600,000 and 1 million were slaughtered, or died on the marches. Other tens of thousands fled north, to the relative safety of the Russian Caucasus. Hundreds of thousands of women and children, we now know, were compelled to convert to Islam and survived in the families of Kurds, Turks, and Arabs.[17]

The deportees who reached the desert areas of Syria were abandoned in the eastern Zor province, in makeshift camps, without food, decimated by epidemics and further massacres; there, they largely perished. Franz Werfel wrote that millions of people had also been deported from the war fronts of Europe, and yet

> Where the evacuees of the European theater had been led from the battle zones for their own protection, had been cared for even in hostile territory, and never lost hope in being allowed to return to their homes after a bitter but not unexpected hiatus, the Armenians could expect no protection, no help, no hope. They had not fallen into the hands of an enemy who, on reciprocal basis, had to respect international law. They had fallen into the hands of a far more terrible, unfettered enemy – their own country.[18]

Defeat in the Balkan War, the 1914–18 war, the revanchist movement led by Kemal and the Treaty of Lausanne brought profound upheavals for people of Greek origin, who had settled in the territories of the Ottoman Empire for millennia. At the beginning of the war, the empire counted, in the territories of present-day Turkey, almost 1.8 million subjects of Greek nationality (including many Greek Muslim converts in central Anatolia); by the end of the 1920s, less than 200,000 remained in the Turkish republic, the majority in Istanbul. Various figures have been conjectured, in the order of several hundred thousands, about the number of victims of the anti-Greek massacres and persecutions, especially in the Pontus region. Historically, apart from the large community in Istanbul, Greeks were concentrated in Izmir and in the Pontus region along the eastern Black Sea coast. The government's nationalist policy for the Turkization of Anatolia had already before the outbreak of war intimidated and forced some 150,000 Greeks to leave the coastal provinces and take refuge in the Aegean islands close to the coast. In 1915, numerous contingents of Greeks were deported from the coast to the Anatolian interior.[19] In 1921–2, with the defeat and retreat of the Greek army, a large section of the Greek population, fearful of Turkish reprisals, left the country; its number is estimated at 400,000–500,000. A further 250,000 Greeks living in Eastern Thrace, which Kemal had reconquered, returned to their homeland:

> After the Turkish victory [over the Greeks], two groups of Greek Orthodox were left in Asia: the communities in central Anatolia, which were partly Turkish-speaking (the so-called 'Karamanlis') and those living on the Eastern Black Sea coast, the Pontian Greeks. Each of these groups numbered about 200,000 souls. In addition there were several hundred thousand Greek Orthodox in the Constantinople area, partly residents of long standing, partly refugees.[20]

During the negotiations for the Treaty of Lausanne, an agreement was signed between Greece and Turkey to exchange Muslim citizens living in Greece and Orthodox citizens living in Turkey (the agreement specified religion, but not language). The final outcome of this exchange was 1.3 million returns to Greece and 300,000 returns to Turkey. These were returns of a forced nature, accepted not only by the two contracting countries, but also by the international community and the League of Nations. With this exchange, the Turkization of Anatolia (and its European appendix) had been completed: whereas before the war, minorities counted for one-fifth of the population, by 1923 their share had fallen to less than 3 per cent. Killings, flight and deportations had rendered the new republic solidly homogeneous, with a 'Turkish' identity, freed from religious affiliation.

New and radical forms of forced migration appeared with the end of the Ottoman Empire. As compared to the many cases in the past, here forced migration arose from a political ideology born out of complex and numerous motives. At the forefront of these was the fear of the final disintegration of the empire and of a secular order in which ethnic and religious diversity were tolerated. Defeat in the Balkan War, the first disastrous year of the Great War and the bitter defeat suffered at the hands of Russia in the Caucasus, the simultaneous threatening attack by the Allies in the Dardanelles, and the heightened sense of danger, all reinforced nationalist ideology. The ruling group believed that a new Turkey should be based on a homogeneous and loyal Turkish-speaking society, free of heterogeneous minorities. Only Muslim minorities, such as the Kurds, or those composed of Turkish-speaking refugees, could be 'Turkized'. The other minorities were included in a plan of 'demographic engineering' (supported by legislative decrees and based on solid bureaucratic and administrative data) that provided for the deportation and dispossession of the Armenian minority, and was accompanied by violence and massacres, to the point of slipping into genocide. This violent process would end, in a less bloody but still forcible manner, with the population exchange decided in 1923.

The deportation processes concerned entire population groups and were thus not selective in strictly demographic terms. However, although the overwhelming majority of Armenians and a good proportion of Greeks were peasants, the minorities disproportionally included members of the country's cultural, economic, commercial and cosmopolitan elite. Finally, the close link between ethnic cleansing through deportations and genocide represented an ominous portent of the tragic events of the 1930s and 1940s.

2.4. The Soviet Union and its internal enemies

According to the calculations of nonprisoner engineers, six people can sit on the bottom bunks of a Stolypin compartment,[21] and another three can lie on the middle ones (which are joined in one continuous bunk, except for the space cut out beside the door for climbing up and getting down), and two more can lie on the baggage shelves above. Now if, in addition to these eleven, eleven more are pushed into the compartment (the last of whom are shoved out of the way of the door by the jailers' boots as they shut it), then this will constitute a normal complement for a Stolypin prisoners' compartment. Two huddle, half-sitting, on each of the upper baggage shelves; another five lie on the joined middle level ... and this leaves thirteen down below: five sit on each of the bunks and three are in the aisle between their legs. Somewhere, mixed up with the people, on the people and under the people, are their belongings. And that is how they sit, their crossed legs wedged beneath them, day after day. (Aleksandr Solzhenitsyn)[22]

Stalin had been dead for twenty years already when Solzhenitsyn introduced the world to the workings of the Gulag archipelago, the carceral system of forced labour camps, places of confinement and internal exile. In and around these locations unfolded the forced migrations of entire ethnic groups, uprooted from their settlements and forcibly relocated to remote parts of the Soviet Union. During the war 'From Petropavlovsk (in Kazakhstan) to Karaganda, a Stolypin car might be seven days en route (with twenty-five people in a compartment). From Karaganda to Sverdlovsk it could be eight days (with up to twenty-six in a compartment).'[23] As we shall see, forced migration, together with the wartime 'evacuations' and the great migrations driven by the (also forced) industrialization decreed by Stalin at the end of the New Economic Policy (NEP) in 1929, were the main axes of mobility on the Soviet continent. The possibility of uprooting millions of people from their historical places of settlement against their will relied on three conditions. The first was that there needed to be clear political determination – something the Soviet regime certainly did not lack, despite contradictions and reversals. The second was that this determination had to be translatable into concrete actions, without causing unrest, chaos and uncontrollable conflicts. Here, the 'secular arm' of the regime, the People's Commissariat for Internal Affairs (NKVD), provided the instrument of efficient organization and repression. The third was the existence of a logistical system adequate for the massive movements of people that took place over time. Useful, to this end, were both the extensive railway network which spanned the USSR's vast territory, and the available experience of the management of mass transfers, dating back to Tsarist times. Finally, the great expanse of the country beyond the Urals, Siberia, the Far North and the Far East offered space for settlements and opportunities for agricultural, industrial and mining developments.

A general idea of internal migration movements between the 1926 and 1939 censuses (the period marked by dekulakization[24] and the great famine of 1932–3), can be deduced from Lorimer's seminal study.[25] In the 1926–39 period, European Russia (excluding the Ural region, Bashkiria and Dagestan) had a net loss owing to emigration of some 5.4 million people, and Kazakhstan almost 1 million. By contrast, there was a net immigration of 6.4 million people into the Transcaucasus (1.4 million); the Ural region, Bashkiria, Siberia and the Far East (3.3 million); and Central Asia (1.7 million). The USSR's urban centres absorbed more than 25 million people from rural areas.[26] This redistribution took place both through more or less spontaneous movements, as well as through planned ones stimulated and supported by the Stalinist policy of industrialization. It also comprised forced migrations, mainly consisting of the deportations of the kulaks, the 'rich' peasants, which involved millions of people.[27] In the 1920s and 1930s, Soviet policy was above all oriented towards redistributing labour. 'Millions of people moved between the Soviet republics as the state attempted to regulate internal migration primarily by stimulating resettlement to sparsely populated regions with considerable deposits of natural resources, including to northern and far-eastern Russia and to Kazakhstan.'[28] In order to regulate internal migration, to control the population and dissenters, and to regulate and slow down the urbanization processes, a system of registration of the population (over 16 years of age) residing in towns, workers' collectives, *sovkhoz* and construction complexes was introduced in 1932. People in these categories had to have a passport, registered by the police, in order to obtain a residence permit, or *propiska*.[29] This system did not, however, prevent other more or less free, more or less voluntary, migrations from taking place. 'The industrial labor force, especially in such fields as lumbering, mining, industrial construction, and railway, canal, and road building was augmented by the assignment of prisoners, deported kulaks, and those condemned for political deviations to various undertakings, especially in remote and relatively undeveloped areas.'[30] The Soviet Union's entry into the war of course saw the strengthening of the many restrictions on internal mobility, but also an explosion in other forms of mobility related to the war situation. It is worth mentioning, first of all, the massive evacuation of civilian populations in 1941 and 1942 under the hammer blows of the German invasion. It is estimated that 16.5 million people were evacuated to the rear, away from the front, to (what were presumed to be) safe areas. Rejecting the traditional term 'refugee' – suggesting a more or less voluntary choice to move – the term 'evacuee' was coined, 'whereby civilians would be designated by the state for displacement and transferred in an organized fashion to the rear, where they would become productive participants in the war effort'.[31] This meant avoiding, through organized mobilization, the disorder that would come with masses of 'spontaneous' refugees. With the throng of evacuees, the workforce also moved, together with plant, machinery and raw materials and other resources saved from enemy clutches.

'[E]vacuation came to resemble another form of population displacement with which Soviet authorities were increasingly well acquainted, namely deportation', including dekulakization and the removal of 'enemy' ethnic groups, which had already begun in the 1930s. But very often things did not run as planned: 'Masses of people were on the move, travelling without documents and in conditions in which it proved difficult to corroborate their identity . . . Indeed, in the eyes of the authorities, the evacuation had provided new cover for enemy agents, spies and saboteurs.'[32] At the end of the war – it should be mentioned in passing – not everyone was able to return to their homes; those who did not return were mainly those who had found stable employment in the districts where they had been resettled.

The travails of politics and war both drove the Soviet population's internal mobility and interfered with it (meanwhile, its external mobility was reduced to practically nothing). During the war period, there were ever-more episodes of forced migration of ethnic minorities who were deemed unreliable (or 'fifth columnists'), though there had also been numerous precedents for this. In fact, the first of the three waves of deportation of the Ingrian Finns (18,000 in 1929–31, 10,000 in 1938, 20,000 in 1941),[33] whose loyalty to the new regime was doubted, dated back as far as 1929. The 200,000 or so Chinese who had lived in the far east of the Russian Empire in the 1910s were practically eliminated by coerced repatriations, as well as the escapes, the executions and the very high mortality in the forced labour camps. As early as 1926, a plan had been outlined for the deportation of the Korean minority living in the far eastern regions bordering Korea, which was suspected of having reached understandings with the Japanese occupier of that country. However, the deportation was enacted only in 1937, when 172,000 Koreans were deported to Kazakhstan and Kyrgyzstan, in areas close to the Aral Sea. Left without material or financial support, they were decimated by the soaring mortality rate. Those who survived were never allowed to return to their areas of origin.

With the USSR's entry into the war and the German invasion, there were growing numbers of deportations of the ethnic groups deemed liable to make accommodations with the aggressors, or who lived in territories occupied by the advancing Germans or had collaborated with them, often only for survival's sake. Even in those instances where members of a given ethnic group had enrolled in the German army, these were in all cases much lower in number than those from the same ethnic group who fought in Red Army ranks. Among the first to be deported were 58,000 Greeks from Pontus, who had resided in Ukraine for centuries and were transported to various parts of Russia and the Caucasus in three waves, in 1942, 1944 and 1949, as 'politically undesirable elements'; they were only allowed to return to their residences after 1956. In 1943, it was the turn of 496,000 Chechens and Ingush, who were sent to Kazakhstan and Kyrgyzstan, and to the Russian Soviet Federative Socialist Republic (RSFSR), where they were kept in labour camps and 'special settlements' under NKVD control. In Kazakhstan, Kyrgyzstan and Central Asia, 70,000 Circassians (1943) and 8,000 Balkars (1944) were deported; 95,000 Mesketi Turks were sent to Central Asia (1944). For 191,000 Crimean Tatars (deported in 1944) their destination was Uzbekistan and, for 93,000 Kalmyks, Siberia.[34]

These deportations had certain traits in common: the absence or near-absence of advance notice; the length of the train journeys; the poor food, hygiene and safety conditions during and after the journey; placement in labour camps and 'special settlements' with extremely harsh living conditions; the absence of basic infrastructure; and extreme climatic conditions. In each case there was also a very high mortality rate in the early stages of settlement (one–two years) that some estimates – it is not known how well-founded – put at around 20 per cent.[35] It was only after the Communist Party's

Twentieth Congress in 1956 and Khrushchev's denunciations that the deportees were allowed to return to their homelands (with, as we shall see, the exception of the Mesketi Turks and the Germans). Even this happened only gradually and partially. But the most remarkable element was that the deportations involved entire ethnic communities, including both those who were considered 'politically undesirable' and the hundreds of thousands of men (and women) who had loyally served in the Red Army. Hence, there was no demographic or social selection, here: it is true that women far outnumbered men, but this owed to the fact that these latter had been recruited to the army or died in its ranks.

The largest ethno-linguistic group subjected to deportation were the Germans. Settled in Russia since the eighteenth century, the 1897 census had counted some 1.8 million of them, though this number declined to a bit less than 1.5 million by 1939. Between 3 September 1941 and 1 January 1942, 799,459 Germans were confined to internal areas in 'special settlements'. The number of these *Russlanddeutschen* increased to 1,209,430 at the end of March 1942, including 203,796 who were repatriated from Germany.[36] This accounted for some 82 per cent of the population that had been counted in 1939, and consisted of residents of the lower Volga regions (the Autonomous Soviet Socialist Republic, ASSR, of the Volga Germans, abolished in 1941), Saratov, Stavropol, as well as Moscow, Leningrad and other large cities in European Russia. Their destinations were Kazakhstan, the Far North and Siberia. Yet even in 1939, the Volga Germans, mostly peasants in the *kolkhozes*, had been officially considered loyal and reliable citizens.[37] The outbreak of war abruptly changed that situation; the Kremlin's decree announcing their deportation, published on 8 September, stated that according to the military authorities there were thousands, indeed tens of thousands of 'deviationists and spies' among the Volga Germans, ready to commit acts of sabotage at a sign from the enemy command. In view of this eventuality

– and in order to prevent massacres and the inevitable reprisals – 'the presidium of the Supreme Council of the USSR has found it necessary to resettle the entire population of the Volga regions, under the condition that the re-settlers are allotted land and given state aid to settle in new regions'.[38] After the end of the war, the Germans were prevented from returning to their original residences: whereas other groups retained their cultural and ethnic characteristics, this did not happen for the Germans, whose diaspora integrated into Soviet society.[39]

The war also caused massive flows of refugees across borders, both in and out of the USSR, in response to the events of the conflict; it further caused millions more forced migrants in the postwar period as a result of the redrawing of the country's borders. Two million Soviets migrated to the territories transferred from Poland to the USSR, and numerous other transfers occurred along the entire European border. But these movements were the consequence of the catastrophe of the war, rather than a political plan to interfere with and manipulate mobility.

Lenin and Stalin accepted the multi-ethnicity of the state, opposing the creation of a Russian nation-state: 'they nevertheless accepted its principle and tried to create the basic foundations – a territory, elites, a language, a culture – for each Soviet ethnic minority'. This, paradoxically, made it easier to uproot and deport those ethnic groups – especially ones close to the state borders – suspected of sympathies that crossed borders.

In the 1930s and 1940s, the state was a powerful driver of migration movements, largely standing above individuals' choices or indeed those of the communities or ethnic groups to which they belonged. The various figures cited above are, for the most part, indicative in nature: they underestimate the movements that escaped scrutiny or otherwise eluded statistical accounting, as well as the many refugees who perished during the transfers or whose traces were lost. They mix together reasonably orderly transfers and brutal deportations; destinations relatively close to the places of residence and very distant ones at the other end of the world. We see something like a spectrum of situations, ranging from essentially free movements to incentivized ones, from the movements made compulsory by complex regulations to forced ones – in the worst cases meaning actual imprisonment. The totalitarian state's power made all kinds of manipulation of mobility possible.

III
Misdeeds of Nature

3.1. Unkind nature

Don't be fooled by the subheading: nature is not, in itself, unkind. Rather, it has played a paramount role in shaping mankind's spread across the planet, and hence in the migrations that have tended to populate the most suitable regions in terms of climate, natural resources and communications. Humans have done their best to adapt to the most diverse natural situations, settling the planet from the Arctic Circle to the Equator, including even marginal and precarious regions. In these and other areas, nature has sometimes been 'unkind', due to the stress situations created by climatic variations, or by shifts in the Earth's crust, or by changes in the universe of microbes. Many of these stress situations have created drives to migratory movements, set in motion by the natural instinct to 'flee' situations that endanger survival, thus depopulating the affected regions. These are forced migrations, set in motion not by the decisions of a tyrannical state, but by a state of necessity. They are migrations that would not have occurred without the occurrence of a particular event or natural condition. There exist various conditions in which natural stress can bring about migration; but here we will not deal with temporary migratory movements in order to escape some immediate danger, followed by a return once the danger had receded. Common throughout human history has been the migration-escape to get away from the contagion of an epidemic (fleeing the infected city, like the young men and women in the Decameron, just as today people move away from regions stricken by COVID-19), only to promptly return once the emergency has passed. Or to get away from an erupting volcano, only to quickly return home once the danger is over (indeed the area around Mount Vesuvius is one of the most densely populated in Italy!).

This chapter will discuss natural traumas that have generated long-term migrations, either because they have changed environmental conditions or because they have upset delicate balances. Today, there is much discussion about whether global warming causes or accelerates processes of desertification and generates – domestic or international – migratory flows of rural populations to other regions. We will then speak of 'climate refugees' and of 'climate migration' – which, it can be presumed, is rather diluted across time. This chapter deals with well-identified natural phenomena that have had lasting effects on the mobility of vast regions. One such case was the episode of prolonged drought known as the 'Grande Seca' (great drought) of 1877–9 in the state of Ceará and in the so-called 'drought polygon' in north-eastern Brazil. It opened up new outlets for migration, or resumed ancient routes with mass migration movements. Another was the 'Dust Bowl' of the 1930s, which drove hundreds of thousands of migrants out of Oklahoma, Texas and Arkansas due to a renewed and prolonged drought. These are specific, well-documented episodes. Great droughts have also been blamed for the decline and disappearance of the Maya civilization in Yucatán and Central America towards the end of the first millennium AD, or the abandonment of the settlements of the Pueblo Indians in the southwest of today's United States at the end of the thirteenth century. These are only hypotheses about phenomena with uncertain contours, and nor is it certain that it was climate events that caused the eventual emigration flows.

At the opposite pole are the catastrophic consequences of floods, cyclones, earthquakes and tidal waves, which have scarred certain populations, breaking delicate balances or making long-established settlements impracticable and risky. The tormented history of Haiti, hit by the tremendous earthquake in 2010 and the devastating Hurricane Matthew in 2016, is related here as a reminder of the vulnerability of poor and overpopulated lands such as this Caribbean island country, as well as the flow of emigration it provoked, from Haiti to Brazil, and then to Chile and Peru and finally to the United States, following old migratory routes and opening up new ones. Central and South America periodically suffer climatic upheavals due to the El Niño cycle, with major repercussions on settlements and migrations.

The third example is due to the natural cycle of a parasite: the blight, which for five years affected and destroyed potato crops, the staple food of the people of Ireland. There is a close and unequivocal link between this parasite, food, poverty and emigration; the blight was the fuse that set off the famine, making it untenable for millions of people to stay on the island. The parasite destroyed the harvests of a society that was already suffering from a very serious Malthusian imbalance – certainly also fed by distant political factors. It was, however, a natural event that, in a chain reaction, changed the migration that had already been directed towards Great Britain and America for decades, from a moderate flow to an unstoppable exodus.

The effect of natural factors – of nature's misdeeds – on settlement and migration has, over time, been strongly mediated by social and economic factors. Their effect is much higher on rural populations than on urban ones, and in fragile overpopulated areas (the aforementioned Haiti, or the low-lying coasts of southeast Asia hit by the 2004 tsunami) and less organized societies. Some of these natural events are cyclical (for instance, the periodicity of El Niño) and to a certain extent predictable, and thus give populations the opportunity to equip themselves to mitigate their effects; others are unforeseen and strike at unprepared societies. In the history of migration, the relations between humans and nature have had a powerful importance that cannot be overlooked; some of their pathological aspects will be mentioned in these pages.

3.2. Drought

He contemplates the ruin of the *fazenda*; ghostly oxen, alive no one knows how, fallen under the dead trees, their skeletal legs stumbling in the dried-up bushes, they wander aimlessly, swaying; an ox that has been dead for days but remains intact is shunned even by the urubus, for their beaks do not penetrate its dried-up skin; jururu oxen go in search of the clearing, in the brushwood-covered ground, where their favourite puddle used to be; and – the thing that aggrieves him most – those oxen which are still not completely exhausted, seeking him out, confidently surround him, mooing a long sad call that sounds like a cry ... The northeasterly wind continues intense and impetuous, whistling across the plateau in plaintive heartbreaking notes among the dry branches of the caatinga, and the sun rises, reverberating the unquenchable fires of the heatwave in the clear firmament. The man of the *sertão*, overwhelmed by misfortunes, finally bends ... One day the first herd of fleeing *retirantes* passes by his door; astonished, he watches them cross the fields, miserable, and disappear in a cloud of dust at the bend in the road. Another day passes. And then another. It is the *sertão* that empties out.

He can no longer resist. He joins one of these groups, the paths cluttered with skeletons, and proceeds on the painful exodus towards the coast, or the distant mountains, or anywhere where the primordial element of life is not lacking (Euclides da Cunha).[1]

This is how Euclides da Cunha describes the effects of drought in the *sertão*, the inland region of north-eastern Brazil periodically affected by this misfortune. An equatorial semi-arid region, it was populated by farmers, labourers and breeders, whose subsistence depended on the scarce and irregular rains. This was a population dispersed over large expanses, with a strong ethnic mix, in a region where the first seeds of development were slow to appear. Recurring droughts quickly translated into resource scarcity or outright famine, and these caused malnutrition, disease, mortality and migration. In Ceará, a state half the size of Italy, with a population of just over 700,000 at the time of the 1872 census, the Grande Seca hit hard in the three years 1877–9. The people of Cebrano still had memories of the drought of 1844–5, which had occurred a generation before, and the deaths and migrations it had brought. The chronicles bear witness to other severe droughts of the past, in 1724, 1778 and 1792, 1817 and 1827; and other similar, now well-documented catastrophes occurred in 1915 and 1932 and, more recently, in 2012.[2]

The Grande Seca had Ceará as its epicentre; there are no precise statistics, but some reliable estimates indicate 120,000 Ceará people being forced to emigrate to Amazonas state, and 68,000 to other destinations. This emigration took place despite initial resistance from the authorities and the big landowners, who worried about the loss of manpower. The state capital, Fortaleza, was then a town of just 25,000 inhabitants, but 114,000 poor and hungry *retirantes* (as the fugitive emigrants were called) poured into it, amassing in public spaces, along the coast, and in the immediate vicinity of the town, alarming local residents and creating numerous problems of public order. Seeking to contain the pressure, the authorities organized and financed the transportation of the *retirantes* to São Paulo state, in the far south, but especially to the neighbouring states, as well as to Pará and Amazonas. The *retirantes* sought employment mainly in agriculture, public works and mining. The authorities also set up a series of farming colonies in Amazonas state, in order to boost its meagre agricultural production, which had been bled dry by competition from a rubber industry whose rapid growth had sucked in manpower during these same years.[3] Hundreds of *retirantes* from Ceará were employed in the first attempt to build the railway designed to connect the valleys of the Mamoré with that of the Madeira, the longest tributary on the right bank of the Amazon. The route was meant to stretch 400 kilometres from Porto Velho to Guajará-Mirim (towns that emerged together with the railway), on the border with Bolivia, in the middle of the jungle, which caused great difficulties for its construction. The railway was supposed to serve the development of the rubber economy, but the project failed, and construction was abandoned.[4]

The demographic cost of the drought consisted not only in mass emigration, but in a very high mortality rate owing to the outbreak of a devastating smallpox epidemic; according to one local doctor and scholar, in two months of 1878, there were 24,212 smallpox deaths in Fortaleza – 9,721 in November and 14,491 in December.[5]

The recurring droughts in north-eastern Brazil – a region of more than one million square kilometres – were constantly accompanied by emigration flows, some of them permanent. These were, certainly, internal migrations within Brazil; and yet the capital Ceará stood thousands of kilometres from São Paulo in the south or Manaus in the west, destinations that represented different worlds for the *retirantes* fleeing the Grande Seca. The natural shock was a powerful drive to migration.

The Brazilian Grande Seca of the 1870s and the Dust Bowl of the 1930s in the Great Plains of North America are separated by six decades, by several thousand kilometres, and by the long cycle of development between a primitive and isolated agriculture and a modern, highly mechanized one embedded within a large economic system. These situations would thus seem incomparable, if not for the fact that similar climatic factors created similar migration flows.

The Dust Bowl migrations occurred in the 1930s in the southern part of the Great Plains (in Canada called the Prairies), and particularly in Oklahoma, Texas, Arkansas and Colorado.[6] The Great Plains had a semi-arid climate, very cold winters, very hot summers, and low and irregular rainfall.

> [T]his ecoregion was dominated by short- and mixed-grass prairie vegetation prior to European settlement. Between the US Civil War and the start of the 1930s, approximately 30% of the US portion of the Great Plains was converted to cropland, with much of the remaining grassland used for livestock grazing. Agricultural settlement developed a few decades later on the Canadian Prairies than in the US, but similar land use patterns had emerged there as well by 1930. In both countries, governments had policies that encouraged the establishment of family-operated farms on the Great Plains through a process known as homesteading.[7] Although a number of fast-growing urban centers had developed on the Plains by the 1930s, the population remained disproportionately rural, with local livelihoods and regional economic systems tied strongly to agriculture.[8]

The growing expanse of cultivated land was used for wheat, maize, cotton and soya crops, which required deep ploughing. It took place without rotations, with the destruction of the turf, and it generated soil drying and erosion. In times of drought, the dried-out soil was pulverized and frequent and widespread dust storms were generated: the 'Black Sunday' of 14 April 1935, when dust-laden clouds obscured the daylight, would remain infamous. The years 1931 to 1939 were all drought years, with peaks in 1934 and 1936; sand and dust storms were very frequent, as was widely documented in the films and photos of the time. The migratory movements linked to the Dust Bowl were the consequence of exceptional climatic events, but so, too, the result of the collapse of a precarious agricultural system, based on crops unsuited to the environment, in an economic system severely depressed by the great crisis of 1929. The high prices of wheat, maize and cotton in the 1920s had attracted strong immigration by farmers – many of them inexperienced – and stimulated mechanization and production. With the great crisis, the prices and incomes of debt-ridden farmers plummeted, and they were forced to produce more to meet costs and debts, or to abandon their farms and emigrate.[9]

The crisis linked to the Dust Bowl and its aftermath caused heavy emigration domestically and a decline in the populations of certain states: Oklahoma's fell from 2.4 million in 1930 to 2 million in 1945, while numbers dropped from 1.9 to 1.7 million in Kansas and from 1.9 to 1.8 million in Arkansas.[10] California, the state that attracted the largest share of migrants, increased its population from 5.7 million to 9.3 million during this same 15-year period.

The Dust Bowl migration was turned into a modern epic by newspapers, films and novels,[11] which constructed a narrative that did not respect the complex reality of the facts. Of the 300,000–400,000 migrants who left Oklahoma and neighbouring states, not all of them made their way to California. Only one-third of them stayed in the valleys of the interior and found work in agriculture, mostly as labourers, while the majority sought work in the Los Angeles area.[12] This was not an emigration of farmers alone: a good half of the emigrants came from urban centres or worked in trade and services, which were also victims of a national crisis aggravated by the collapse of agricultural activities in the region. Interestingly enough, the wave of Okies, Arkies and Texies (given these disparaging names in accordance with their origins in Oklahoma, Arkansas and Texas) arriving in California and the other destination areas was rather poorly received, despite the fact that these migrants were white, Protestant and of British origin.[13]

The Brazilian and American migratory crises, which occurred in such different contexts, call for reflection. The climatic crises – the droughts – must be seen in relation to the human factors that helped create them: the land system, cultivation techniques, the price regime and public policies. In the Great Plains, the patterns of land occupation, human settlement and production were – as we would say today – not sustainable for long; society became vulnerable and entered a serious crisis when the extraordinary climatic events of the 1930s hit. Similar considerations could be made regarding the Grande Seca and the recurring droughts in north-eastern Brazil: an extremely concentrated distribution of land, with forms of production vulnerable to climatic extremes, i.e. livestock farming, large plantations and small farmers reliant on subsistence production. In both cases, these were extremely scattered, recently settled populations whose institutions, techniques and ways of life had not adapted to the environmental circumstances.

3.3. A Caribbean odyssey

Not having heeded the advice of Admiral [Columbus], these noblemen set sail from the river and the city of Santo Domingo. And once the fleet had set sail, it happened that eight or ten leagues from departure, they ran into such bad weather, that out of thirty ships and caravels, four or five were saved, and the greater part of them were shipwrecked along the coast and many were swallowed up by the sea, because all trace of them was lost. And more than five hundred men drowned... many gentlemen, hidalgos and men of valour. And it was there that that nugget of gold was lost, which, as I have related, weighed more than three thousand six hundred pesos, and more than one hundred thousand pesos of gold and many other things, so that it was a great loss and a cursed day. (Gonzalo Fernández de Oviedo)[14]

This account by Fernández de Oviedo, *cronista oficial de las Indias*, refers to the fate of the great fleet that had brought the new governor to Hispaniola; upon its return journey to Spain, it was wrecked by one of the violent hurricanes typical of the Caribbean region. The chronicles of the West Indies are replete with reports of the catastrophic consequences of hurricanes sweeping through the region in summer and autumn; the one reported by the *cronista* destroyed the Spanish fleet in early July 1502. Ten years earlier, Columbus had landed on the northern coast of the island (then baptized Hispaniola, today divided into the Dominican Republic and Haiti) where Môle-Saint-Nicolas currently stands. The island was home to a couple of hundred thousand Taino, an ethnic group that first wretched exploitation and then disease drove to extinction in just over half a century. The importing of slaves from Africa and the introduction of sugar-cane plantations built a new demographic and economic base for Hispaniola. In 1804, when the Haitian part of the island won its independence (from the French, who had taken over from the Spanish), the population, almost entirely of African descent, numbered half a million inhabitants, growing to over three million in 1950 and to 11 million in 2020.

Haitians inhabit a land of 27,000 square kilometres (about the size of Massachusetts). Hemmed in by a largely mountainous territory, despoiled by its gradual deforestation in the bid to create new farmland, their population density is equal to that of the Netherlands (with over 400 inhabitants per square kilometre). Deforestation has a long history, beginning with the introduction and spread of the sugar-cane plantations in the seventeenth and eighteenth centuries; with the export of valuable timber such as mahogany; and with the introduction of unsuitable models of industrial agriculture. But since the 1940s–1950s, this process has accelerated: population growth (the number of Haitians has quadrupled since 1940) has pushed the rural population to expand the spaces under cultivation, moving up the hillsides and mountains and deforesting them, and gradually eroding the wooded cover. While in the 1940s woodland blanketed more than half of the island's surface, today that has been reduced to 30 per cent. Coal production for domestic use has aggravated this problem. Soils have become vulnerable to flooding, which has caused further erosion and loss of fertility. The uncontrolled use of pesticides has led to serious water pollution problems, especially in the valleys and lowland areas. Like the other Caribbean island states, Haiti is regularly plagued by hurricanes and tropical storms (2004, 2008 and 2016 being the most disastrous)[15] that devastate a fragile environment, with heavy damage to crops, and the destruction of housing and infrastructure.

Haiti is a textbook case of Malthusian overpopulation, not moderated by reproductive 'prudence', and with the highest mortality rate in the region.[16] Migratory exodus meanwhile has relieved some of the demographic pressure: in 2020 1.8 million people born in Haiti lived abroad (the migratory stock), more than triple the half million they numbered in 1990. The main destinations are the United States (where almost half of the Haitian emigrants live), the neighbouring Dominican Republic (about a quarter), Canada, Brazil and France. Net emigration between 1990 and 2020 is estimated at almost one million.[17] Emigration – for many Haitians and many other poor populations in the Caribbean and Central America – has been the most efficient way out of poverty both for those who have managed to leave and for those who remained in their homelands. Migrants' remittances (worth over a third of GDP in 2019) go directly into the pockets of poor people and not into those of bureaucrats, middlemen or the underworld, as has too often been the case with development aid.

The close intertwining of growing demographic pressure and multidimensional poverty (Haiti has the lowest per capita income in the Western hemisphere, the lowest education levels and the highest mortality rate) prevents us from isolating the net effect of environmental degradation on the migration phenomenon. But nature – in addition to hurricanes and tropical storms – hits the island with tremendous force, with unforeseen consequences on emigration. On the early evening of 12 January 2010, an earthquake (registering 7 on the Richter scale) devastated Haiti with terrible destructive power. The epicentre was only 25 kilometres from Port-au-Prince, and there was tremendous devastation and a very high victim count: over 300,000, according to a hasty government estimate, which subsequent investigations reduced to much lower figures between 100,000 and 200,000. The material damage to dwellings, production facilities and infrastructure was enormous, estimated at between one and two times Haiti's annual GDP. Some 300,000 homes were destroyed and still in late 2010, an estimated 1.3 million people were displaced in emergency camps.[18] The disaster pushed the Haitian diaspora in a new and unexpected direction. Brazil, which had already engaged personnel in Haiti in 2004, as the head of the United Nations' Stabilization Mission,[19] became the destination for a new flow of immigration. At first, the flow followed a very bumpy route, first to the Dominican Republic by bus, then by plane to Panama, then by land, air and sea through Colombia, Ecuador and Peru to the border with the Brazilian state of Acre, in the Amazon region.[20] The arrival of tens of thousands of Haitians in a poor, backward region with no reception facilities led the state of Acre to declare a situation of 'social emergency' in 2013, before organizing the forced transfer of the migrants to São Paulo. Meanwhile, the Brazilian government's officially declared opening to Haitian immigration allowed for direct entry into the country. The new flows were directed towards the southern states: Santa Catarina, Rio Grande do Sul and, above all, São Paulo. Haitians found employment in the construction and meat-processing industries, settling in the interior areas of these three states rather than in the big cities. Official statistics speak of 67,000 residence permits, both temporary and permanent, granted up until 2016. According to other administrative sources, there were 98,000[21] entries in 2014 and 2015 alone. The state of Acre reported 130,000 entries at the state border after 2010. With the economic crisis and the fall of Dilma Rousseff's government in 2016, the gates closed again, and an influx of departures from the country, to Chile and to the north of the continent, began, in the hope of reaching the United States and Canada. In January 2010, the United States approved a 'Temporary Protected Status' (TPS) for people fleeing the disaster on the island through which several tens of thousands of Haitians found refuge. But in 2017, Donald Trump declared the programme terminated; those in the US would have to repatriate or emigrate elsewhere by 22 July 2019: this affected about 59,000 people. Other extensions were later granted, but the doors to regular immigration were effectively closed again. Until the next catastrophe? Thus, after meandering throughout the continent, the Haitians' odyssey ended at the US border. Mexican sources estimated that 30,000 Haitians entered Mexico by heading north through the Guatemalan border; by the end of 2016, 12,000 migrants had arrived in Tijuana, at the US border, but could not cross it, causing a humanitarian crisis. Given the porousness of the borders – and the spread of the informal economy everywhere across the Ibero-American continent – many Haitians had been dispersed along the lengthy route from Brazil to the United States.

The Haitian case is an example of the disorder of the world – and the hardships and suffering that this imposes on migrants. Borders that open and close, flows that are first welcomed and then rebuffed, migrants without rights, and on-off humanitarian interventions, in an international climate in which the verb 'close' is uttered much more often than 'open'. But it is also an interesting case of a migratory flow that takes a new course

– namely, towards Brazil – after a natural disaster. Now this flow has dried up. But it is possible that the path traced over the last decade will be trodden again, should migration policies change.

The climatic turbulence typical of the Caribbean has also tormented Central America – and indeed, continues to do so. These disturbances are linked to the El Niño cycle, the periodic climatic phenomenon that causes the warming of the waters of the Central and South Pacific adjacent to Latin America, which results in alternating violent floods and droughts. In 1998, Hurricane Mitch caused an estimated 10,000 to 20,000 deaths in Honduras and Nicaragua, but events of the same nature, if not the same magnitude, occur annually. In concluding these pages, dedicated to the consequences of an earthquake, it is worth returning to the climatic drives to migration. Approximately 45 million inhabitants live in turbulent Central America, south of Mexico; there is a very marked socio-economic fault line separating them from Mexican society, indeed similar to the one between Mexico and the USA. Mexico has a GDP per capita more than double that of bordering Guatemala and near-neighbours El Salvador and Honduras (the so-called 'northern triangle'). This fault line is not just economic, since it is deepened by the violence and insecurity generated by the extremely high crime and bloody gang warfare, more devastating even than that ravaging Mexico.[22] In the 2014–18 period, the crisis was aggravated by a prolonged and very serious drought that severely impoverished rural populations, exacerbating the push factors for northward migration. In one fortnight in 2020, the two hurricanes Iota and Eta raged, causing hundreds of casualties and hundreds of thousands of displaced persons, and the destruction of crops and harvests.[23] The migratory flow from the northern triangle, joined by smaller flows from other Central and South American countries, the Caribbean and even Africa, is certainly not caused by the adverse weather events alone, but they have exacerbated poverty, particularly that of the four million farmers living in the so-called dry corridor, the Pacific belt stretching from Nicaragua to Mexico.[24]

Until 2018, it was relatively easy to get into Mexico via the Guatemalan border, by means of rather adventurous journeys, especially for those arriving from more remote regions. Migrants were given a transit permit valid for sixty days, after which they had to return home. The migrant caravans travelled the 3,000 or 4,000 kilometres between the Guatemalan border and the northern border cities (Tijuana, Mexicali, Ciudad Juárez, Nuevo Laredo) by various means: buses, trucks and freight trains. Many of these transits took place aboard the 'Beast',[25] one of the many goods trains that rumbles through Mexico from south to north, with frequent stops, line changes and interruptions. The authorities, for the most part, did not intervene. The journeys are long, uncomfortable and risky; dangerous because of the numerous cases of accidents, thefts, extortion, violence and ill-treatment, not to mention the abuse suffered at the hands of the traffickers (*polleros, coyotes*) who organize the illegal border crossings. Reliable sources estimate that each year these transits involve between 300,000 and 400,000 people. In many cases, these are repeated transits by people who have already been rejected or deported, who attempt the adventure one more time. It is by this route that the Central American community in the United States has gradually grown, now numbering more than 3.5 million migrants. However, the growing number of transits is now dammed up by the closure of the US border, which makes the irregular flows even more risky and difficult, in a dead-end funnel. In border cities (such as Tijuana and others) the number of Central American (and other) migrants waiting for an opportunity to enter the USA (the now-rare acceptance of an asylum or humanitarian-protection request, or else clandestine transit) is growing. Many find precarious jobs in the informal economy. Others – a few for now – go back to their homelands.

The two cases discussed – the aftermath of the Haitian

earthquake and drought in the northern triangle – are very different, even though they concern the same region of the globe. Without the earthquake, there would not have been a new route for Haitian migrants. But if the drought had been less severe, the emigration flows from the dry corridor would probably have taken place just the same, perhaps in smaller numbers. Both cases confirm the paradigm that the negative effects of adverse natural events are exacerbated by environmental, social and economic fragility, in turn affecting migration. Climate change owing to global warming can only aggravate the vulnerability of Caribbean and Central American populations, influencing migratory pressures. This is not to say, however, that adequate and costly social and economic development interventions cannot contain these pressures in these critical regions of the planet.

3.4. Ireland: the blight of diaspora

No country in the world, in the history of the world, has endured the hemorrhage which this island endured over a period of a few years for so many of her sons and daughters. These sons and daughters are scattered throughout the world, and they give this small island a family of millions upon millions who are scattered all over the globe, who have been among the best and most loyal citizens of the countries that they have gone to, but have also kept a special place in their memories, in many cases their ancestral memory, of this green and misty island. So, in a sense, all of them who visit Ireland come home.

Most countries send out oil or iron, steel or gold or some other crop, but Ireland has had only one export and that is its people. (John F. Kennedy)[26]

When John F. Kennedy began his triumphant state visit to Ireland in 1963, the island – including the province of Ulster –

counted only half the eight and a half million inhabitants it had upon the eve of the Great Famine, in 1845. This represented a historic low, a unique case in modern history: a country whose population halved in the space of little over a century. Ireland was the source of a diaspora that began at the end of the eighteenth century, intensified in the thirty years following the end of the Napoleonic Wars, and became an unstoppable torrent in the century that followed the Great Famine. The first American census, in 1790, counted some 150,000 inhabitants of Irish origin – a number which has now reached 30 million.[27] The rapid acceleration of emigration around the midpoint of the nineteenth century is inextricably linked to a natural development, namely the spread of potato blight, a micro-organism that for five years, between 1845 and 1850, relentlessly destroyed the Irish population's main staple crop.[28]

The link between the potato blight and emigration may at first seem a little puzzling, like the theory that traces the decline of the Roman Empire back to the poisoned water in lead pipes.[29] But such a conclusion would be mistaken: for history and scholarly research have provided abundant proof of this connection. The Famine was the event that burst the dam containing a population that had tripled in the space of a century.[30] This was a very poor population, living in semi-serfdom, in a limited territory, with a density similar to wealthy regions like the Netherlands or Lombardy. The reasons for its demographic dynamism were owed to a systemic change: 'in the late eighteenth and early nineteenth centuries it is clear that the Irish were insistently urged and tempted to marry early: the wretchedness and hopelessness of their living conditions, their improvident temperament, the unattractiveness of remaining single, perhaps the persuasion of their spiritual leaders, all acted in this direction'.[31] Rural populations had neither the possibility nor the habit of saving money, and the idea – common to many European populations – of putting off marriage until a small nest egg was available to start a family was totally alien to them. Big landlords tended to keep tenants at subsistence level, playing with the lever of rents and hindering improvements in living standards. Marriage was not itself costly; the house, usually little more than a hut, required only a few days' work with the cooperation of friends and family; furniture was simple and rudimentary. The fundamental problem, in a society of tenant farmers, was the availability of land for new households to settle. As long as this remained difficult (except in the case of inheritance, for instance upon the death of the family father), marriage was discouraged. But towards the end of the eighteenth century, conditions changed. The conversion of pastureland to crops and the cultivation of new land (whether reclaimed, mountainous or marginal), promoted by the reforms of the Irish parliament, as well as Britain's demand for agricultural produce during its war with France, reduced this obstacle. The subdivision and parcellation of land increased, favoured by a new development – the extensive spread of the potato as the main and often almost exclusive food of the Irish population. The special role of the potato, perhaps introduced by Sir Walter Raleigh at the end of the sixteenth century and then gradually established, was decisively owed to its high productivity. With the population becoming increasingly dependent on potato consumption, 'land which formerly had been adequate for only one family's subsistence could be parceled out among sons or other subtenants'[32] since 'an acre of potatoes sufficed to feed a family of six and the livestock'.[33] In addition, the potato, which has good nutritional value, was consumed in incredible proportions, as part of an everyday diet that also included a considerable consumption of milk.[34] Thus, the availability of new land and the subdivision of existing land, made more productive by potato cultivation, enabled the high nuptiality and low marriage age of the Irish. This, combined with high natural fertility and not particularly high mortality, sustained the high rate of population increase in the period before the Great Famine.

This system was not destined to last for long and already in the 1830s there is evidence of a slight increase in the age at marriage; emigration to America was already beginning to take on larger dimensions. In the decade 1825–34 the average annual number of departures from Ireland was 35,000, a number which increased to 45,000 in the following decade:

> The typical emigrant up to 1830 and perhaps 1835 was a small famer, often impoverished and ill versed in his own business, but proud of his independence and determined to improve upon it in the new world. Such men often had to work as unskilled laborers when they arrived in America, but possessing a knowledge of English and aided frequently by the presence of friends in the country, those with energy had little difficulty in rising to better jobs.[35]

However, those who emigrated later on were labourers with no skills except that of growing potatoes on their own patch, 'hampered by ignorance of the land and of the language, and by a character in which excesses of joy and gloom seemed equally unfortunate to the slower tempered Anglo Saxon'.[36]

It is plausible to think that this change in the demographic system would have developed gradually, if it had not been for the occurrence of a tragic event. So, let us return to the blight; already in summer 1845 there were reports of the loss of the potato crop in southern England; in the following month the blight appeared in Ireland and in October it was found that between a third and a half of the harvest had been lost. The crisis worsened in 1846, with three-quarters of the crop ruined; in the following year – later called 'Black '47' – the blight spared the crops, but as very little was sown, very little was harvested. In 1848 and 1849 the blight struck again, causing heavy losses, and it was not until 1850 that things returned to normal. These figures do not themselves tell of the enormity of the catastrophe caused by the parasite: the population was reduced to starvation, which directly killed an unspecified number of people; typical malnutrition-related diseases such as typhus and various types of fever spread; cholera also made its appearance and mortality increased three or four times over. It has been estimated that the Great Famine, and associated epidemics, caused between 1.1 and 1.5 million excess deaths.[37] The British government implemented contradictory and ineffective policies: early attempts were made to support the poorer parts of the population with public works, but the programmes were suspended. Soup kitchens were set up all over the country, but these were not supported at adequate levels, for fear of creating a sort of popular dependency on public funds. The government ended up deciding that responsibility for aid to the most unfortunate strata of the population should be offloaded onto local communities through the workings of the Poor Law. The large landowners drove hundreds of thousands of impoverished tenant farmers off their land, but continued to export grain to Great Britain, where it fetched favourable prices.

The catastrophe shook the economic and social system to its foundations, and decimated the population, which fell from 8.2 million in 1841 to 6.5 million in 1851, due to the combined effects of mortality and emigration. This was the last resort, at least for those who could afford to pay their way, and were not wearied by age, hunger and disease. Already in 1845, some 75,000 Irish people left the ports of the United Kingdom for America, as compared to 54,000 the previous year; in 1846 they numbered over 100,000, and in Black '47 itself there were 215,000 of them; the vessels that transported them from Liverpool, or Dublin or Cork or other ports, were called 'coffin ships' because of their characteristic crowded conditions, filth, disease and high mortality rates. Over the next seven years, up to the end of 1853, the annual exodus varied between a minimum of 150,000 and a maximum of 250,000, before then falling to 79,000 in 1855. Thus, across the whole decade of

1845–54, some 1.8 million Irish people set course for America, more than a fifth of the island's population as counted in 1841.[38] In the following decades, up to the end of the century and beyond, the flow of emigrants remained very substantial, with Ireland losing between one and two per cent of its population each year.[39] Emigration remained an option for economic and social redemption for what was then northern Europe's most backward population.

It should also be added that the option of migrating became gradually less difficult and expensive as the century progressed. In 1840, the Cunard shipping line put four steamships into regular transatlantic service, but this was limited to an expensive passenger service; from the 1850s onwards, steamships began to transport emigrants, and by 1860 they had become the almost exclusive means of transport. However, even in the first two-thirds of the nineteenth century, difficult, precarious, irregular crossings had managed to transport millions of people to the other side of the ocean. Much progress had also been made in the age of sailing ships; the British government repeatedly enacted *Passenger's Acts* – in 1803, 1823, 1825, 1828, 1835 and 1840 – to regulate the various aspects of transatlantic navigation, from the characteristics of the vessels to their equipment, to regulations for ship-owners and passengers.[40] The boats became larger, safer and faster; the sailing ships of the first decades carried few migrants, at most a few hundred; the large clippers that entered service in the 1840s carried between eight hundred and a thousand migrants, with better travelling conditions. As already mentioned, from the 1850s onwards, sailing vessels were rapidly abandoned in favour of modern steamships; the travelling time dropped from over ten weeks to two; travelling conditions improved enormously; and migrants also faced fewer hardships in the days before embarkation and after arrival. Moreover, the formation and growth of Irish colonies in America made things easier for newcomers, who found relatives, countrymen, friends and solidarity there.

In short, emigrating became easier, at the same time as the demand for labour on the other side of the Atlantic was growing rapidly.

Without doubt, the catastrophe of the Great Famine broke out due to a natural factor – a micro-organism that spread from its American habitat to Europe. The Great Famine threw the island's entire economic and social system into crisis and – some would argue – fed its rebellion against Britain. The years of the Famine prompted between a fifth and a quarter of the island's population to take the route of mass emigration: a route that had already been open and travelled for more than half a century, but in much smaller numbers. After the Great Famine, emigration became the permanent means of restoring the social, economic and demographic balance. But not only that. For the demographic regime, based on a low age at marriage and the universality, or near-universality, of marriage, was also revolutionized. The shock of the Famine did not affect the reproductivity of married couples: this remained very high, because the various forms of voluntary birth control and contraceptive methods spread much later in Ireland than in the rest of Europe. But it did affect the 'frequency' of marriage: people married much later (the average age at first marriage for women rose from 23–24 before the Great Famine to 27–28 towards the end of the century), while many (one in five) men and women remained unmarried.[41] The new demographic regime based on emigration and low nuptiality was supported by the big landowners and the clergy, and produced a sharp population decline. The four and a half million-strong Irish population counted in 1901 amounted to little over half of the numbers of sixty years hence.

IV
Organized Migration

4.1. On the road, not alone

Migration is often costly, uncertain, risky. The calculation of costs and benefits is complicated, and individuals minded to migrate often lack the necessary information to make a decision. This is, indeed, almost always the case, unless the decision to move is determined by a state of necessity, such as extreme poverty, hunger or imminent danger. In such cases, the benefit is in saving one's life, and there is no cost that can offset it. In the modern world, many migratory movements have occurred because they have been organized by institutions standing outside the family, the clan or the local community. They might be organized by a lord, a corporation, a religious order or a state, which generally benefits from this population displacement, whether it is outgoing or incoming. We can define such displacement as 'organized migration'; it encourages the migrant to take this decision, assuring him of certain benefits and providing a sort of insurance against the risks of displacement. In truth, the organization of these displacements takes different forms and varies in intensity. But those who promote them always have a certain intention, whether it is economic, political or even demographic in character. The organization of the means of travel also functions as a drive or accelerator of mobility processes: we saw one example of this in chapter I, with reference to the founding of Greek colonies in the Mediterranean.

This chapter explores three very different cases of organized migration. The first concerns the young women, the so-called *filles du roi*, sent from France to Quebec, Canada. This migratory movement was limited in numbers but very significant in terms of the motivations behind it and its results. It involved less than a thousand young women from charitable institutions and homes, sent to the nascent French colony within the course of a few years and settled in the vast territories of the Saint Lawrence valley. They had been earmarked to marry settlers or garrison soldiers as soon as possible – and, above all, to have children. The journey, the trousseaus and dowries, and their upkeep while awaiting marriage, were all funded from the coffers of Colbert and Louis XIV. The motivation for this costly initiative was, first and foremost, of a demographic nature and – as we shall see – in this respect it was highly successful.

The second case concerns a period of three centuries, and the considerable masses of migrants who made up the *Drang nach Osten*. This 'drive to the East' was part of the German expansion beyond the Rhine and Elbe and towards the vast territories inhabited by scattered Vend populations. This was a movement organized and financed by princes, bishops and religious orders, and it resulted in the founding of thousands of communities and the cultivation of large tracts of land using more advanced, more productive techniques. It took the form of a migratory movement facilitated by the authorities who organized it, with substantial benefits for the migrants themselves. Counting hundreds of thousands of migrants in the first waves, and tens of millions of descendants, this produced the Germanization of vast territories.

The third case – more circumscribed, both in time and in its mass of numbers – concerns the immigration of tens of thousands of German settlers into the vast territories of the lower Volga valley, as per the project desired and implemented by Catherine the Great in the 1760s. This project financed the colonists' journeys from the Rhine valley and their settlement beside the Volga, providing them with tools, livestock and seeds, and granting them vast plots of land which they now owned. This experiment was too costly to be continued, but was successful both in terms of the economic development of the region and in demographic terms, due to the very rapid growth of immigrant communities. Catherine the Great's experiment was part of the movement begun with Peter the Great, accelerated from the 1760s onwards, of intensifying the peopling of the southern territories of European Russia, controlling semi-nomadic ethnic groups, and Russifying the territories taken from the Ottoman Empire. This policy was also followed by Maria Theresa to strengthen the Danubian borders of the territories ceded by Turkey.

Absolute states in the seventeenth and eighteenth centuries were highly amenable to mercantilist doctrines and the principle that populous territories and growing populations were the basis of the wealth and strength of nations. They were similarly seen as functional for the defence of the nation's territory. During the consolidation of Prussia, the settlement of colonists in the new territories – in Silesia after it was taken from Austria, but also elsewhere – was strongly supported by Frederick the Great. There were other attempts at organized migration into depopulated territories. But the efforts made by the Lorraine dynasty in the Maremma of the Grand Duchy of Tuscany, or of Charles III in Andalusia, quickly ended in failure, to the great detriment of the migrants themselves.

The three cases presented here are examples of migration policies supported and organized by powerful political authorities. These powers could influence the 'fitness' of migrants and the success or failure of their relocation. History, however, serves up a wide variety of ways in which migratory movements were organized, up to the example of the 1950s and 1960s, when the strongest countries pursued programmes to recruit foreign migrants to support the reconstruction and expansion of their postwar economies. In the 1940s, the United States recruited Mexican labourers to replace men called up to the armed forces in carrying out seasonal work in the Southwestern countryside; in the 1950s Belgium signed conventions with Italy to recruit workers for its mines; until the 1960s Australia had agencies in Europe to feed the flow of European migrants, keeping at bay a feared influx of migrants from Asia.

4.2. The *filles du roi* in the laboratory of *Nouvelle France*

> From the most recent letters that I have received from Canada, I have been made aware that the girls who were transported there last year, having been taken from the General Hospital, were not hardy enough to withstand the climate and the cultivation of the land, and that it would be more advantageous to send young country-girls who are in a condition to endure the hardships that must be faced in that country. And since these can be found in the parishes around Rouen ... I have thought that you will find it appropriate for me to beg you to use the authority and credit that you have over the curates of thirty or forty of those parishes, to find in each of them one or two girls who are willing to travel voluntarily to Canada to settle there.[1]

When Colbert wrote this letter, the Canada to which he referred was the Saint Lawrence valley region, with a stronghold, Quebec, that was little more than a village, and two other small settlements, Montreal and Trois-Rivières, with a few hundred souls, plus sparse houses and farms scattered over a vast territory. In the era of Colbert (1619–83), and Louis XIV, France was Europe's most populous country, with over 20 million inhabitants, while Quebec had 7,000 French settlers and an indigenous population a few tens of thousands strong.[2] The British colonies were more populated, but still had a small population.[3] They were also far away, and their contacts with the French colonists were limited to mercantile relations and local conflicts over territories and trade routes. From the founding of Quebec in 1608, with the first stable core group of colonists settled by Samuel de Champlain, up until 1663, when the colony passed from the rule of the private company known as the 'Hundred Associates' to direct control from Paris, immigration had been very modest. According to the 1666 census, there were 3,215 settlers, two-thirds of whom were men; they were mostly hunters and fur traders, soldiers, clergymen and *travailleurs engagés*.[4] Immigration accelerated from 1663, with the arrival of Intendant Talon: while between the founding of Quebec in 1608 and 1660, there were about 5,000 immigrants, in the following period up to 1700 their number almost doubled. This was a modest flow, next to nothing compared to the French population of the time (just eight emigrants per year for every million French people); conversely, neighbouring Britain, with one-third as many inhabitants, produced a net immigration to New England of 280,000 people between 1630 and 1700.[5] Rigorous research has established that only one-third of those who immigrated before 1700 (4,997 migrants) went on to start families: the others either returned home, died before marrying or remained there (in very small numbers) single or unmarried. The biological 'founders', i.e. those who started their families in Quebec (very small numbers were already married at the time of immigration; others, the vast majority, married after arrival), were the founders from whom descended the 70,000 inhabitants whom this territory could boast in 1760, when it was conquered by the British. But immigration had not added to this increase: after 1700, it had fallen to a trickle and in many years, there was greater *emigration*.[6]

France had recently emerged victorious from the War of Devolution with Spain but making only meagre territorial gains, and Louis XIV was strongly committed to consolidating his primacy in Europe. The powerful minister Colbert was a staunch supporter of the idea that a flourishing population was the foundation of the national wealth and was reluctant to commit human and financial resources to support emigration flows that could weaken France's demography. In addition, there was a widespread belief that France was going through a period of demographic decline. Mercantilist ideas permeated the political culture of the time; 'With rare exceptions [economists] were enthusiastic about "populousness" and rapid increase in numbers ... A numerous and increasing population was the most important *symptom* of wealth; it was the chief *cause* of wealth, it *was* wealth itself – the greatest asset for any nation to have.'[7] When Jean Talon,[8] the energetic intendant of Nouvelle France, asked Colbert to give a push to immigration, he replied: 'It is not befitting [the king's] prudence to depopulate his kingdom such as would be necessary to do to populate Canada ... the country will populate little by little and, with the passing of a reasonable time ... [the population] will become very considerable.'[9] No mass migration, then; after all, the subject of the French crown was forbidden from emigrating except in certain conditions, and necessarily to the colonies. These had to grow, above all, according to their natural dynamics. But, as already mentioned, it was unthinkable that a population of a few thousand inhabitants, predominantly males, could reach any significant size. Colbert realized that the demographic and economic development of the new colony would have no few benefits for the mother country. On 5 April 1669, Colbert – a progenitor of modern demographic policies – extended to Canada the provisions already issued for France three years earlier:

> His Majesty has ordained and commands that in the future the inhabitants of the said country [Canada] who have a number of at least ten living children, born in lawful wedlock, and who are neither priests, nor men or women of the cloth, shall be paid from the funds which he shall send to the said country, a pension of 300 livres for each year, and to those who shall have twelve, 400 livres.[10]

In addition, men who married before the age of twenty were granted a 'king's gift' of 20 livres on their wedding day; it was also stipulated that a fine was to be imposed on parents whose children did not marry before the age of twenty, if they were male, or sixteen, if they were female. The state coffers were in no danger, as all were married off at a very young age; moreover, despite very high reproduction rates, it was very rare that at any given time parents would have ten or twelve living children, all of whom had escaped the appalling levels of infant and juvenile mortality. However, these provisions are indicative of the climate of the time.

One boost – limited in numbers, but effective in its results – came from the patronage that Louis XIV gave, from 1663 to 1673, to the emigration to Quebec of young women, who were either single or widowed. They would later be called the *filles du roi*.[11] Via the religious and administrative authorities, in a decade, about 800 young women were selected and sent across the Atlantic – a small number, but equal to half of all the women who immigrated to Canada before 1680. They were picked out from charitable institutions for orphans and the destitute as well as from mostly very modest families in Paris, the Île-de-France region and north-western France. Each migrant woman cost the king 100 livres: 10 for the recruiter, 30 for the purchase of personal effects and 60 for the sea crossing;

the new arrivals were housed and fed in the early days after their arrival, until marriage or other arrangements could be made, and many of them were given a dowry at the time they were wedded. They were young women, with an average age of about 24; very few of them were widows, nine out of ten were illiterate and half were fatherless. It was said that theirs was a free decision; it can be assumed that the prospect of a life in New France could appeal to orphaned, illiterate and mostly very poor girls, without the need to impose the decision over their own wishes. Once they arrived, they married quickly: 8 per cent within the first month, almost 40 per cent within three months, and 83 per cent within six months. There were even official 'edicts' that bachelors had to choose a bride within 15 days of the ship's arrival . . . in order to avoid penalties such as a ban on hunting in the woods. The rapid marriage of the *filles du roi* and of the few young women available for wedlock was aided by the stationing of the 1,200-strong Salière-Carignan Regiment, sent to contain the Iroquois threat to the colony. Many of its members would permanently settle there: several hundred soldiers decided, at the end of their service, to remain in Canada, taking advantage of substantial incentives and land concessions. But even without them, the number of marriageable men was several times the number of available young women, who had no shortage of opportunities to find a spouse.

Despite the small numbers, and the insignificant contribution that immigration made to population numbers after 1700, the French colony grew rapidly, increasing fivefold between 1700 and 1760. This owed to factors that we will go on to discuss later. The fact is that immigration brought migrants to Canada who had undergone a strong biodemographic selection. According to Intendant Talon, only individuals 'who are of the right age for procreation and above all those who are truly healthy' were to immigrate to Canada – those who were not lame, and who did not have recognized diseases. As far as women were concerned:

It is most opportune that it be strongly recommended that those who will be destined for this country are not in any way malformed by nature [*disgraciées*], that they do not have anything visibly repulsive, that they are healthy and strong for work in the countryside, or at least that they have some capacity for manual labour.[12]

Another brutal selection was made by the harshness and length of the journey, on average lasting several weeks. During this voyage, a not insignificant proportion of the passengers succumbed.[13] This 'bio-physical' selection combined with the environmental conditions of the destination itself. Its climate was surely harsh, but salubrious and not too different from that of northern France, from whence the settlers hailed. There was ample space and, presumably, less incidence of epidemic diseases; there were abundant natural resources and plenty of food. At least on paper, there was also a certain selection of a social nature, excluding individuals considered undesirable. In short, the particular case of New France – microscopic when compared to the great migratory movements that took place at the time – allows us to verify the outcome of such selection processes. During the eighteenth century, the population increased at the quite exceptional rate of 2.5 per cent annually (equivalent to a doubling every 28 years), while the population of their country of origin was at a standstill; reproductivity was very high, one of the highest ever recorded in the world; the age at marriage low; survival better than in France. Translated into numbers, it is interesting to compare certain demographic data concerning Quebec (1608–1760) and the north-western region of France (from whence the settlers came, 1670–1769). For example, the total number of children born to women married before the age of 20 was 11.8 in Canada and 9.5 in France; the fertility rate at age 30 (for women married at ages 20–24) 479 per thousand versus 403, and the interval between successive births 1.8 years versus 2.3. Moreover, women in

Quebec married at an average 2–3 years younger than their counterparts in France.[14] The natural population increase was consequently very high, ten times that of France in the same period, five times that of the average European population.

This migration to Quebec was an extreme case of migration that was not only 'organized', but also aimed at populating a territory. Even with minimal effort, the mother country had achieved a massive success. Since the effort was sustained for too short a time, this did not pay off politically – Quebec would, after all, change hands. France, in reality, had not transported settlers, but 'reproducers' – almost a successful laboratory experiment.

4.3. The *Drang nach Osten* and the Germanization of Eastern Europe

> As the land was without inhabitants, he sent messengers into all parts, namely to Flanders and Holland, to Utrecht, Westphalia and Frisia, proclaiming that whosoever were in straits for lack of fields should come with their families and receive a very good land – a spacious land, rich in crops, abounding in fish and flesh and exceeding good pasturage.[15]

This was how Helmold, in his *Chronica Slavorum*, reported the proclamation made by Count Adolf of Schauenburg, lord of war-torn Holstein, around 1140. This appeal was not destined to fall on deaf ears; indeed, the Germanic expansion, the *Drang nach Osten* (drive to the East)[16] beyond the boundary set by the river Elbe, had long since been set in motion. This expansion led to the Germanization of vast territories; it was also a political process, supported by complex migratory flows which colonized the less populated territories to the east which had hitherto been settled by Slavic peoples.[17]

This great movement began in the eleventh century and gained pace over the next two centuries; its foundational phase would come to an end with the crisis of the fourteenth century. The means of its development are highly instructive; indeed, it is worth dwelling on them further, because these migrations followed a pattern that could not be repeated in the crowded modern world. To simplify things rather, we can say that this process unfolded along three main axes: the southern one, along the natural course of the Danube, towards the plains of Hungary; the middle one, in the open territories from the Netherlands to Thuringia, Saxony and Silesia, north of the central Bohemian mountains; and finally, the northernmost one, which avoided the swampy territories and forests of Germany, which rendered migration and settlement difficult, and instead ran along the Baltic coast, bit by bit leading to the foundation of cities such as Rostock and Königsberg. Slavic settlement was thus pushed back to the east and massively penetrated by migration into Austria in the south, Silesia in the centre, and Pomerania and Prussia in the north. Beyond these areas, which nevertheless continued to varying degrees to have Slavic ethnic groups, the – albeit fragmentary – migration continued and penetrated into the Baltic provinces, Volinia, Ukraine, Transylvania, Hungary and even further east.

This gradual colonization process bears some resemblance to the wave of advancement within Europe that took place a few thousand years earlier, which developed from the south-eastern Mediterranean in a north-westerly direction towards the British Isles.[18] That was a slow advance, whose operation presumably consisted of spontaneous settlements by successive generations each in search of new land.[19] In the case of Germanic migratory movements – and these are the ones that interest us, here – this was intentional migration, not caused by expulsion, but rather organized and guided by institutions and elites that combined to produce a real migration policy. At the head of the colonization process were the princes, such as the Margrave of Meissen, the bishops and, later, the knights of the Teutonic Order, who deployed considerable resources to this end. In the eleventh and twelfth centuries, colonization developed beyond the line formed by the rivers Elbe and Saale, which marked the border of the Carolingian Empire and the eastern limit of Germanic settlement. During the twelfth century, Holstein, Mecklenburg and Brandenburg were colonized. In the thirteenth century, immigration peaked in eastern Brandenburg, Pomerania, Silesia and northern Moravia, crossing the Oder line. Settlement in Prussia, crossing the Vistula, reached its peak in the fourteenth century. Germanic expansion did not penetrate Bohemia, the interior of Pomerania or Lusatia. The process of eastward expansion continued even after the abandonment of land and the reversals caused by the long demographic crisis of the fourteenth and fifteenth centuries. The *time* of the colonization process is effectively punctuated by the process of founding new towns, which reached its peak around 1300.

The emigration to the East, the *Drang nach Osten*, was the largest of the great medieval colonization processes, but not the only one. Under the impetus of these movements, the European continent assumed a very stable settlement structure, which the depopulation process following the period of the plague – visible in the large number of abandoned villages – depressed but did not subvert. Much of the European space was occupied by farmsteads, villages, castles and towns, in a network of population centres that proved both stable and resilient. The medieval migration movement to which we refer here raises many questions which are highly important to understanding this subject. The first concerns the dimensions of the migration phenomenon, which we have to estimate largely on the basis of conjectural elements. The study of the various available forms of documentation – censuses, land registers, the founding of villages, etc. – has led to the hypothesis that in the twelfth century the migratory phenomenon from old Germany to the territories between the Elbe and the Oder involved around 200,000 people.[20] In the subsequent century, a similarly sized movement colonized further territories, reaching as far as Pomerania and Silesia. The documentation indicates that in Silesia, between 1200 and 1360, some 1,200 villages were founded and other 1,400 in East Prussia, amounting to a total of 60,000 farms and a population that can be estimated at around 300,000 people.[21] These are, of course, relatively small figures, but they are to be compared to a modestly sized Germanic population amounting to a few million (around six million in 1200). Even supposing that the estimates are far below the real figures and that the migration flows were two or three times higher, this would still mean very modest levels of migration, not exceeding one per thousand annually. However, it must be considered that this relatively small flow had a conspicuous 'founder' effect (few progenitors, many descendants), when we consider that at the end of the nineteenth century the Germanic populations east of the Elbe–Saale line approached 30 million.[22]

Some important questions arise here. First of all: can the eastward movement be considered – as is often assumed – a movement driven by a hunger for land? That is, one resulting from the increase in agricultural density in the areas from which migrants departed, itself so high thanks to vigorous demographic growth? There are several question marks over this traditional interpretation, given the still low density of the departure areas (especially at the beginning of the movement); the relative smallness of the emigration flow compared to the vigorous natural growth; and the existence of still relatively unpopulated areas in these territories. The flow of emigration in fact seems to have owed rather more to the high organizational and technical level of the original population and the correspondingly less developed state of the Slavic population in the areas to which they emigrated. Added to this were the very favourable conditions for peasants to settle, and the relatively short distances between the areas of departure and arrival. The German immigrants had ploughs, axes and tools that allowed for the clearing and cultivation of difficult land; the Slavs practised hunting and fishing and itinerant farming, abandoning the fields when their fertility was exhausted. The circumstances and characteristics of this immigration, organized and planned by the clergy and nobility, the knightly orders (Templars and Teutonic Knights) and the great religious orders (Cistercians, Premonstratensians), can be summarized as follows. There was, firstly, the ability to plan and select land – and this was uncultivated land – by measuring it and dividing it, taking into account the availability of water and the risk of flooding. There was also a considerable supply of capital, which was needed to support the emigrants' journeys, to feed them until harvest time, and to provide them with seeds, tools and raw materials. 'The German immigrants were equipped with the heavy, wheeled plough with coulter and mouldboard, and they had heavy felling axes with which to clear the thicker forests in order to cultivate the heavier soils.'[23] Timber was needed to build houses and the village church, and it was also necessary to build the mill and other structures essential for community life. The organization of these migrations required an intermediation between the founding lord and the peasants: there emerged the figure of the *locator*, or 'populator', with real entrepreneurial characteristics.

> The locator contracted with the owner of the land to enlist peasants and to settle them on specified lands. In return for this labour, the contractor usually received a part of the land of the village ... The populator, after recruiting a team of peasants, organised the clearing of the land ... after which the land was measured and subdivided. Accurate measuring was an art that required professional *mensuratores*.[24]

It is not known whether the *locatores*, or populators, also played a selective function when they recruited settlers, although it can be presumed that this did happen. These entrepreneurial and organizational skills allowed the typical family to be assigned a farm (*hufe*) of some twenty hectares (17 hectares, small or Flemish model, 24 hectares, large or Frankish model) with typical settlement of a village of 200–300 people (scattered houses were the exception). Moreover, the land remained free of encumbrances for many years and could be bequeathed, sold or left behind. Contracts between settlers and landowners were regulated by the so-called 'German law', which specified rights and guarantees.

These highly favourable conditions, and the need to send out recruiting agents, suggests that the supply of land was greater than the demand:

> The great extension of the movement is only explained by the fact that colonists bred colonists; for all over the world new settlers have big families. Migration from Old Germany in many cases slackened early. Conditions of tenure in the colonized areas also encouraged this colonization by colonists' families. Law or custom favoured the undivided inheritance of peasant holdings; so there were many younger sons without land.[25]

Going by this approach, we have good reason to doubt the hypothesis of a process set in motion solely by the pressure of strong demographic growth (which surely existed) producing land hunger. Rather, the process appears to have been a self-driven one, encouraged by the abundance of land and the superior organization and technology of the settlers compared to the indigenous populations, which were sparsely settled and pursued a more backward form of agriculture. For the prospective migrants, exchanging a small farm in their land of origin for a property of a few dozen hectares must have been highly attractive. These favourable conditions caused, in turn, a strong demographic growth among the initial core of colonists, which then generated further waves of migration. So, it is possible that this colonization process did not depend on great long-distance migrations, but a continuous march forward, under the impetus of new generations of settlers' children, who became ever more numerous.

We could thus define the *Drang nach Osten* as a slow advancing wave – almost a thousand kilometres from West to East over three centuries – conceptually analogous to that which propagated agriculture from the Middle East to the British Isles a few thousand years earlier. Yet it was different, because of the planned and organized way in which it took place. It was also different because the territories that were gradually occupied were sparsely populated but not deserted. It was different because the economy, techniques, culture and society had, of course, profoundly changed. The German migration to the east was helped along by favourable environmental conditions, the high skills of the migrant farmers, the availability of arable land, the gradualness of the population flow, the low degree of conflict with the native populations and the support of institutions and lords. But its success also owed much to the organizational ability of those who promoted this colonization, based on experience, knowledge of the territory, and a variety of sponsors, from princes and lords to bishops and religious orders. It also owed a lot to the positive overlap between the aspirations of the peasant migrants and the needs of the promoters, between demand and supply of labour and land. In short, this was a 'successful' migration – which would be dearly paid for centuries later, with the two world wars – because it was organized but not coercive, based on empirical knowledge and not on abstract principles, gradual and controlled and not through sudden, unrestrained waves. Such conditions are difficult to replicate, as can be seen from the failure of many absolute states who have attempted to arouse or attract migratory flows without in-depth knowledge of the contexts from which migrants departed and the ones in which they arrived.

4.4. From the Rhine to the Volga with Catherine the Great

> Inasmuch as the vast expanse of Our Empire's territories is fully known to Us, We perceive that, among other things, no small number of such regions still lie unimproved that could be employed with lucrative ease for a most productive settlement and occupation by mankind, most of which regions conceal within their depths an inexhaustible wealth of multifarious precious ores and metals; and since the selfsame [regions] are richly endowed with forests, rivers, seas and oceans convenient for trade, so they are also exceptionally well adapted for the establishment and growth of many types of mills, factories, and various other plants. . . . We hereby solemnly establish, and command that it be implemented, to proclaim it to all. I. We permit all foreigners to enter Our Empire, in order to settle down in any government wherever it may suit each of them. II . . .[26]

Thus begins the manifesto with which Catherine II, in the second year of her reign, opened the borders of the Russian Empire to immigration. The manifesto was made up of ten articles, in addition to this first one, making generous promises to whoever answered the call. It would indeed be answered by many, and overwhelmingly by Germans. This was because the manifesto, which was translated into German and many other languages and actively disseminated through agents in the hire of the Russian government, was most widely circulated in Germany.

At the heart of Europe, in the previous century Germany had been shaken by the Thirty Years' War (1618–48) and the tremendous depopulation that followed. It experienced a reconstruction and recovery process that sparked powerful internal and external migratory movements, which continued even for many decades after the war was over. The human toll of the war and the need for repopulation breathed life into mercantilist theories, according to which a flourishing population was a necessary condition for the wealth of the nation. States and absolute rulers launched repopulation and colonization policies in conquered lands and in sparsely populated border territories, in the effort to strengthen their control. The Habsburg Empire did this in order to consolidate its dominion over the Hungarian and Balkan territories taken from Turkey after the decisive victory in 1683; the Russian Empire did this to populate the large and unpopulated territories to the east and south, and the ones conquered in the conflicts with the Ottoman Empire. Prussia, which was going through its own expansion process, also launched such policies in order to colonize Silesia. It had annexed this region in 1740 and definitively affirmed its control at the end of the Seven Years' War (1756–63), a conflict that had been bitterly fought and come at a vast human cost in the Prussian and Habsburg territories. This same period also saw the onset of mass German emigration to America: the first census in 1790 showed that the German-origin population in the United States had reached close to 300,000.[27] In short, in eighteenth-century Europe, and the Germanic regions in particular, these policies were at the centre of a vast migratory system, which was conditioned, among other things, by changes in these states' territorial structure. This is also the context in which we should place Catherine II's proclamation. A Prussian by birth, bride of the heir to the Russian throne at age 16, and empress by age 32, only a few months prior to this document she had signed the peace treaty with the Habsburg Empire and Prussia.[28] It was followed by an immigration of some 30,000 people between 1764 and 1767, who settled in the near-unpopulated region of the lower Volga valley. This was, therefore, an organized migration, of a kind with the migration policies pursued by the absolute states of the time. The proclamation was issued in all European capitals; many countries placed obstacles or bans on emigration and on the action of recruiting agents, which was effectively limited to the populations of the empire and the Flemish.[29]

Following Article 1, which set out Russia's openness to immigration, Articles 2–5 of the manifesto specified certain bureaucratic formalities, such as the declaration on the choice of activities to be exercised, the choice of place of residence, and the oath of loyalty to the empire. The detailed Article 6 was especially key to the manifesto because it defined the rights and privileges of the immigrant, in particular:

- freedom to profess one's religion and to build churches and places of worship;
- exemption from taxes for varying periods: from a minimum of 5 years for those who chose to reside in St Petersburg and its district or in Moscow, to 10 years for those who enrolled in artisan and merchant guilds, or residents of other cities, and 30 years for farmer-settlers;
- the granting of land to peasants and those who wanted to set up manufactures of various kinds, but without the quantitative indications that were later specified (in the colonial law, Kolonialgesetz, of 1764);
- interest-free loans to farmers for the purchase of livestock and capital goods, to be repaid after twenty years;
- the freedom to import, without duties, goods for personal use;
- administrative self-government for new immigrant communities;
- exemption from military and civil service;
- the freedom for foreign investors to purchase peasants and serfs;
- the payment of board, lodging and travel to the final place of residence;
- the ownership of the land by each community, the prohibition to sell it, and the inheritance of the land by the last-born son, so as to encourage the other sons to pursue other activities.

These were very attractive conditions for people who had suffered the devastation and deprivation of the Seven Years' War. The manifesto was a success, and won over thousands of Germans, mostly from Hesse and the Palatinate, both of which had been particularly hard hit by the war. The immigrants were not only farmers but also artisans and other workers in search of a better life. It should be said right away that the reality was different from the expectations, that the privileges granted to migrants in the early years were not extended to subsequent arrivals, and that Russian openness to immigration was later restricted by various adjustments. The first contingent of 1,600 immigrants arrived from Lübeck, where most of the migrants embarked at the end of summer 1764; the peak was reached in 1766, after a renewed campaign by recruiting agents.[30] After 1767, only a few stragglers arrived: the programme was halted because its operational costs had exceeded the available funds (between five and six million roubles).

The declared freedom to choose one's place of settlement was an empty promise; apart from a few immigrants, considered particularly useful by the government and thus kept in St. Petersburg, almost all the rest of them were forced to join the planned colonies in the lower Volga.[31] Another empty promise was the immigrants' freedom to choose their activity or profession; they were almost all forced to go to the settlements in the Volga and thus become peasants. The sea voyage from Lübeck took nine to ten days; there was a first stop in Kronstadt – the fortified island guarding St Petersburg – and then the ferry to the mainland port of Oranienbaum, not far from the capital.[32] New arrivals were given materials to build makeshift accommodation, as they were held in Oranienbaum for weeks and in some cases months on end before embarking on the long river journey to their final destination. The migrants went up the Neva, crossed Lake Ladoga and Lake Il'men', then began an overland journey of over 300 kilometres to reach the upper Volga. Here, they were again embarked and began a long river journey of almost 1,800 kilometres to their final destination in Saratov. 'When the settlers reached Saratov, they found a disorderly, unattractive and ramshackle community of about 10,000 inhabitants in a region inhabited by Tatars, a river that looked like a sea, and an encampment of shacks and other rough, primitive dwellings.'[33] Within four years, the new arrivals were sorted among the sites designated to house the individual colonies, in a territory that stretched north and south of Saratov on the right and left banks of the river, to a depth of several dozens of kilometres either side of the Volga.[34] The 30,000 settlers were divided into 104 colonies (called 'mother' colonies), each of which was – theoretically – in the centre of a circle 60–70 versts (64–75 kilometres) in circumference, the surface area of which was supposed to be adequate to supporting a thousand families. The colonial law, promulgated in 1764, assigned each family 30 desyatins of land, equal to about 33 hectares of uncultivated land. Each family was also assigned a set of agricultural tools, a wooden plough, two Kalmuk ponies, an ox or a cow, the parts needed to build a wagon, household utensils, and a sum of money from 15 to 25 roubles.[35]

In the initial phase, the mortality (and perhaps the number of escapees?) among the approximately 30,000 settlers who set out for the Saratov region was certainly very high.[36] This ought to be no surprise, given the length of the journey, the harsh climate and the need to acclimatize to a wild region. Moreover, in the early years, drought hit the crops hard, adding to the settlers' hardships. The population declined, to a low of 23,000 in 1773; the crisis was exacerbated by the devastation following Pugachev's Rebellion[37] and raids by Kyrgyz nomads. But the abundance of land, the settlers' capacity for work, and their social cohesion won out; the 'mother' colonies generated 'daughter' colonies, the population increased to 55,000 in 1813, 165,000 in 1861, and almost 400,000 at the end of the century, according to the first imperial census in 1897.[38] This meant a seventeen-fold increase in 124 years (1773–1897) – a doubling every thirty years, i.e. the length of a generation. This increase occurred despite the absence of further immigration, except for the arrival of a few thousand Mennonite Germans, itself more than balanced out by the emigration to America that began in the 1870s. The subsequent history of the Germans in the region became a dramatic one: the Volga colonies became one of the autonomous Soviet Socialist Republics (ASSR of the German Volga) of the Soviet Union in 1924. With the outbreak of the Second World War, the Germans were considered enemies of the state, 'Fifth Columnists' of the Nazi regime. The Volga ASSR was dissolved and all Germans were deported, from 3 to 20 September 1941, in 121 railway convoys to various destinations in Central Asia. Today, their descendants are dispersed in Russia, or have emigrated to the USA, Canada, Argentina and Brazil – or indeed, to Germany, having benefited from the 'law of return' after 1980.[39]

The case of the Volga Germans lends itself to many lines of further investigation, not least as it seems that there is a rich archive of material that has not yet emerged. But some considerations can be advanced already. The first is that despite the difficulties faced, the colonization of this region was successful. One of the objectives was to populate an uninhabited territory, and this was achieved. The 30,000 immigrants had a major 'founder' effect. The abundance of land, family unity, social cohesion and a solid work ethic produced demographic and economic growth, despite many setbacks. The communities' administrative autonomy also presumably had a positive effect. Unlike other cases of organized migration, there was no explicit selection of migrants, although family units were de facto encouraged. The second consideration concerns the effectiveness of the system that was set up: there was a considerable organizational effort on the part of the state, and above all a great economic effort that could not have been replicated later on. This leads to a third consideration, namely that the migratory flow would not have occurred through spontaneous forces, without the powerful funding and incentives – some of which were illusory – that were put in place.

The case of the Volga Germans is emblematic of the solution that the enlightened absolute states found for their hunger for human resources. Russia, from Peter the Great onwards, extended its rule southwards and eastwards, annexing Crimea and other territories taken from the Turks in recurrent conflicts. From 1764 onwards, further impetus was given to efforts to populate the New Russia province, extending from the Bug to the Donets, which formed the southern frontier of the empire. With the defeat of Turkey in the Russo-Turkish War of 1768–74, and with the acquisition of the Crimea, annexed in 1783, the colonization process continued intensively. It was a process driven by Potemkin's government, with migrants from Russia itself but also from abroad, Jews, Greeks, Bulgarians, Poles, Germans and other Westerners.[40] Potemkin also tried to restore the gender balance, which was heavily skewed in favour of men, by forcing soldiers' wives to join their husbands. Moreover, the recruiting agents were given five roubles for every marriageable girl brought to Tauris who married an inhabitant on arrival.[41] At the beginning of the nineteenth century, the incentives for immigration, the mainstay of the empire's southern development policy, came to an end, and immigration dried out, although it had already contributed significantly to population growth. The population of New Russia grew ninefold between 1724 and 1859, rising from 1.6 to 14.5 million.[42]

Not dissimilar was the policy of the Habsburg Empire, aimed at consolidating its hold in border areas with the Turkish Empire, particularly along the Danube from the confluence of the Sava to the Iron Gates. Here, too, success was based on good planning; the wide availability of land; advanced production techniques; the introduction of new crops, such as tobacco and potatoes, and substantial incentives. By the end of Maria Theresa's reign, migration was practically over: almost all the available land had been distributed among the settlers who had established a Germanic-type society on the south-western border of the empire.[43]

The eighteenth century was not just a history of successes; organized immigration also included initiatives that ended in disaster. Some of these concerned the sparsely populated Tuscan Maremma, which had been devastated by the imperial troops deployed by Cosimo I to seize Siena in 1552. The populations had fled and dispersed, reclamation works were abandoned, environmental conditions deteriorated rapidly and malarial fevers spread. Some repopulation projects were unsuccessful; in 1739, Francis I, the first Grand Duke from the Lorraine dynasty, made a more developed effort. A crier was sent to Lorraine, issuing a proclamation in French and German to set out the benefits on offer to colonists; advantages that a later *motu proprio* and other documents allow us to summarize as follows:

> allotment to each family of a bushel of arable land and other land for vineyards, olive groves and vegetable gardens, a pair of working steers, a dairy cow and two sheep, with no obligation to return them for six years, grain for sowing and the necessary rural tools, two pounds and four ounces of bread per day for each head until the settlers have harvested the grains and fodder sown, four pennies per day for each person of any age and sex during the same period, exemption for twenty years from all duties imposed and to be imposed.[44]

It seems that the announcement soon found a positive response that surprised even the authorities; already in summer 1739, the first families arrived, though they had to be kept in Florence, Prato and Pisa before they could move to the Maremma in 1743. The total number of settlers who came to Tuscany – perhaps of the order of 5,000 – is unknown, as many deserted before they were finally settled in the territories of Savona and Massa. Within a few years of its eventual settlement, the colony was already on the verge of extinction, having lost much of its initial numbers. This rapid decline was due both to the abandonment of the colony and to the high mortality rate, a fact well known to the immigration authorities; this mortality was also fuelled by poor hygienic conditions, inadequate housing, abuses in the distribution of food and malarial fevers. In short, the combined effects of slapdash organization, the insalubriousness of the chosen areas and malaria doomed the experiment to failure.

Less disastrous was the Spanish project for the so-called 'new villages of Sierra Morena and Andalusia', with which King Carlos III tasked Pablo de Olavide in 1767.[45] Some 6,000 German, Flemish and Swiss settlers were recruited and settled in fifteen new villages in the unpopulated areas along the Camino Real in Andalusia, with endowments of land, tools, two cows, ten goats and sheep, and a sow for breeding. Again in this case, an Enlightenment idea, common in the second half of the eighteenth century, inspired migration policy: it was supposed that the new colonies, with their particular status, would not bear the negative traits that had caused the decadence of Spanish agriculture, since

> in the new agrarian organisation, the transhumant cattle of the Mesta could not have access, nor could the land be accumulated in a few hands, nor would the regular clergy exercise its pernicious influence, nor would there be room for men of learning and law clerks, except for doctors and administration officials. Problems of all kinds (purely technical ones inherent to the colonisation process, political and financial, and the difficult assimilation of foreigners) caused the relative failure of the experiment, the sole and ultimate cause of which was, perhaps, its improvisation.[46]

The colonies continued to eke out an existence: mortality was very high due to climate change, fevers and other diseases; the gaps were filled in by local peasants, and their particular legal regime was abolished in the following century. With the end of the eighteenth century, the 'organized' migrations pursued by the absolute monarchies of Europe came to an end. In the following century, the territorial structure of the various states became more stable, and their borders took on a definitive, or almost definitive, shape; the boundaries of the Russian Empire had been fixed. The 'hunger' for men eased and then waned, thanks to the demographic transition that accelerated population growth on the continent. Mercantilist doctrines had long since been eclipsed, and Malthus's influence increased. The Industrial Revolution demanded capital and raw materials, and the increase in productivity in the fields and factories meant that rapid population increase was no longer needed. The age of the great transoceanic migration was getting underway.

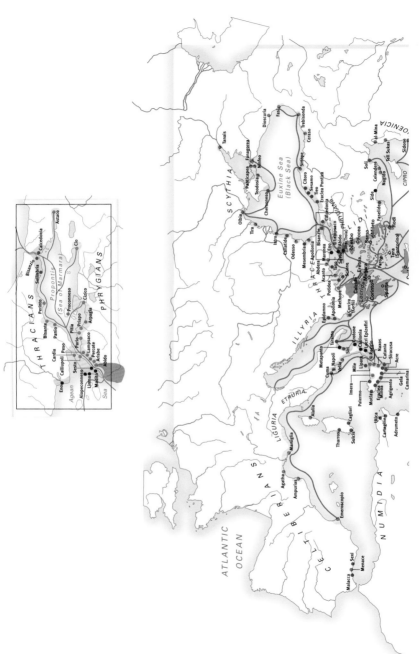

1. Greek colonies in the Mediterranean. © Geo4Map – Novara.

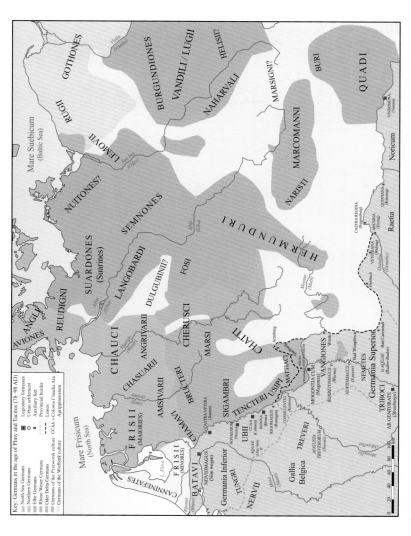

2. Ethnicities of Germania, based on the writings of Pliny (*Naturalis Historia*) and Tacitus (*Germania*), late 1st century AD. Wikimedia Commons.

3. The *Res gestae* of Augustus, fragment of an inscription found in Ankara. Wikimedia Commons.

4. Battle between Goths and Romans, Ludovisi Sarcophagus, circa 251/252 AD, Museo Nazionale Romano, Palazzo Altemps. Wikimedia Commons.

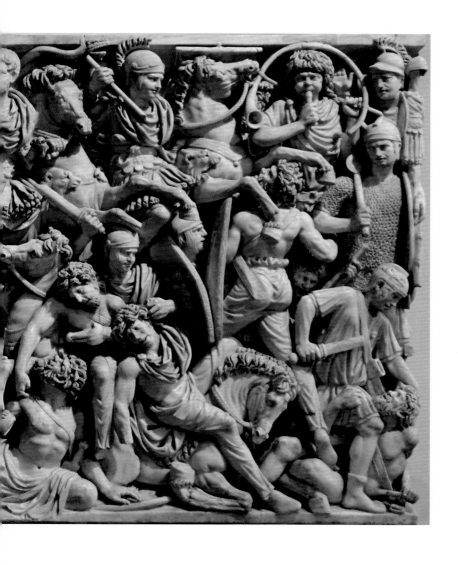

5. The Incas' road system consisted of two parallel routes, one coastal (roughly from the region of Guayaquil to the region of Santiago de Chile) and one mountainous (from Quito in Ecuador to the region of Mendoza in Argentina), each about 5,000 km long, with many cross-connections © Emily Carter.

6. The city of Potosí, Bolivia. Taken from John Ogilby, *America, Being the Latest, and Most Accurate Description of the New World*, London, 1671. The Cerro (mountain) of Potosí provided more than half of the silver destined for Europe, with the employment of thousands of migrants (*mitayos*) who arrived from as far as a thousand kilometres away for the annual *mita* (*corvée*).
© Album / Alamy Stock Photo.

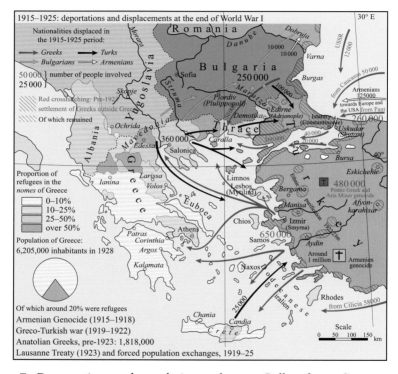

7. Deportations and population exchanges, Balkan front, Greece and Turkey, 1915–1925.

8. Deportation of Germans from Norka, a colony founded in the Volga region in 1767, 60 kilometres south of Saratov, September 1941. 1.2 million Germans were deported between September 1941 and January 1942.

9. Dust Bowl: the arrival of a dust storm. Wikimedia Commons.

10. Port-au-Prince, capital of Haiti, destroyed by the earthquake on 12 January 2010. © REUTERS / Alamy Stock Photo.

11. Allegory of Ireland, Hunger, Emigration and a 'Coffin Ship'. Thomas Nast, 'The Herald of Relief from America', Library of Congress, 91732265.

12. Germanic *Drang nach Osten:* foundation of a village, under the leadership of a *locator*. Illustration from the 'Sachenspiegel' by Eike von Repgow, 13th century (Cpg. 154, Library of the University of Heidelberg). Between the 11th and 14th centuries, thousands of villages were founded east of the line made up by the rivers Elbe and Saale. Wikimedia Commons.

13. The arrival of the *filles du roi* in Quebec, 1667, painting by Charles William Jefferys, Paris, Bibliothéque Nationale de France. Wikimedia Commons.

14. Portrait of Catherine the Great, at the time of the German migration to the Volga region, attributed to Giovanni Battista Lampi the Elder, *c.*1793, private collection. Wikimedia Commons.

15. A transhumant shepherd of the Mesta (a powerful sheep breeders' association) in Spain. Millions of sheep transhumed annually, up to a maximum of 5 million at the end of the 18th century, accompanied by many tens of thousands of shepherds.
© Gianni Dagli Orti/Shutterstock

16. The meeting of the two branches of the Transcontinental Railroad, built by the Central Pacific and Union Pacific, at Promontory Point, Utah, 10 May 1869. Wikimedia Commons.

17. Irish emigrants set off for the United States, from *The Illustrated London News*, 6 July 1850. Between 1845 and 1850, 1.8 million Irish migrants left their country on the 'Coffin Ships'. Wikimedia Commons.

18. The *Titanic* in Southampton harbour. Having set off 10 April 1912, it sank five days later; there were 2,233 people aboard, of whom 1,503 (67%) perished. First class accommodated 325 rich passengers, of whom 125 (38%) perished, but there were also 706 third-class passengers, almost all British, Irish and Scandinavian migrants, of whom 528 (75%) perished. Wikimedia Commons.

19. Italian migrants head to the Opera assistenza emigranti upon arrival in Buenos Aires. According to official statistics, between 1876 and 1930, almost 2.5 million Italian expatriates reached Argentina. © MARKA/Alamy Stock Photo.

V
Free Migration

5.1. A rare phenomenon

What does the adjective 'free' mean, when used to define migration? It is impossible to arrive at any precise definition, if only because this phenomenon is ever-changing. We can content ourselves with an ad hoc definition, suitable for describing, classifying and evaluating migration, and consider as 'free' that movement motivated by an individual decision, which also allows for a choice between 'leaving' and 'staying'. When 'staying' becomes impossible – because it endangers the person's survival or her integrity or produces living conditions she deems intolerable – then we are instead talking about forced displacement, as discussed in chapter II. When 'staying' becomes a less preferable option because some superordinate institution favours, incentivizes and organizes a migratory movement, we are dealing with a situation artificially created by politics (in the broadest sense), which sets on the move a person who otherwise would not have moved. This is why we have chosen to call this type of migration 'organized'. Above, we spoke of the 'individual decision' as a discriminating factor in free migration; but we could quite legitimately also consider 'free' that migration which results from a family decision, or from the shared strategy of a group or community. What has been said does not apply, or applies much less so, to short-range mobility within political boundaries, functional to the community of reference, be it a small rural agglomeration, a village or a township. Here, mobility for family reasons, or for exchange and trade, or for craft specialization, has always been relatively free. But this type of mobility-migration is not our concern in these pages.

However we choose to define it, free migration has been a relatively rare phenomenon in the modern world. It may have been more commonplace in prehistoric times, when spaces were empty, or sparsely populated, and movements were conditioned by predominantly natural or climatic factors. This chapter will look briefly at this type of movement, with reference to the gradual occupation of the European continent. In the modern era, however, such empty or semi-empty spaces have existed only in the Americas; the European world was almost everywhere populated, or at a minimum the empty spaces were under the jurisdiction of a state, a prince or a lord, who regulated access to them.

This chapter examines three examples, three models, of free or semi-free migration. The first of these consists of that migration set in motion by the creation of labour markets able to attract people from across vast distances, requiring a workforce that could not be found locally or in the territories immediately surrounding the areas in need of manpower. In the seventeenth and eighteenth centuries, the – seemingly closed-off and well-established – rural world gave rise to many forms of mobility, mostly of a short-term, recurrent or seasonal nature. This was especially the case where labour in the countryside was free from feudal-type bonds of servitude. There was migration towards large cities, which began to develop in the eighteenth century; to the hubs of maritime trade; to districts rich in manufacturing activities; and indeed to rural areas during peak periods, for harvests or other seasonal work. For the populations, and families, from whom the departees came (very often from mountainous areas), migration supplemented meagre incomes, during a phase of accelerating population growth and increasing pressure on land. Such movement was relatively free of hindrances: there surely were obstacles, but they were not insurmountable, even when the migrant had to cross a border, as was often the case. It was not uncommon for seasonal migrants to marry in the places they came to work, thus preparing the ground for their permanent migration.

The second case is that of the great migration from Europe across the oceans, from the 1840s to the 1920s. The root causes of this migration are well known: this is perhaps the last example of free mass migration, fuelled by two complementary factors. The first was the progressive breaking-down of the constraints on emigration imposed by European states in previous centuries, when they had been dominated by a mercantilist ideology that considered emigration a loss of national wealth and therefore a phenomenon to be opposed or banned. Even in the early part of the nineteenth century, many countries forbade emigration – except on condition of special permits – and in any case those who emigrated lost the nationality of the country of origin and were not allowed to return. But the demographic acceleration of the nineteenth century helped create labour surpluses and added to a poverty that emigration provided a means of alleviating. This itself caused radical changes in states' migration policies. The second factor was the liberalization of entry into the former colonies of the United Kingdom, Spain and Portugal, which in previous centuries had strictly selected immigration flows, mostly reserving entry to their own countrymen. Many countries – Canada, Brazil, Argentina – offered (sometimes illusory) incentives to immigrants. The great migration was basically a free migration, the product of millions of individual decisions. This is, after all, the migratory model that we tend to think about when we consider the migration of the past, although on closer inspection similar phenomena were very rare, regardless of the unparalleled numerical dimensions of these particular movements.

The third case is that of the migratory wave through the North American continent in the 1800s, and the peopling of its territory. This was a wave some 4,800 kilometres long, from the areas closest to the Atlantic seaboard to the Pacific. It was a migration that could not be described as 'internal' because it took place in territories without established jurisdictions, something close to a 'no man's land'. Even if these were not indeed 'no man's lands', the natives (or First Nations, to use the Canadian term) were easily swept aside. Other vast areas of North America were formally French or Mexican territories, but soon passed into the orbit of the United States, either through force of arms or payment. This was, again, a migration surely driven by millions of individual decisions; there was no incentive given by the federal government or the states, and no form of organization other than that which the migrants freely gave themselves. This was an 'advancing wave' without parallel in the Western world, not even in the great expanses of Brazil and Argentina, where the population grew like wildfire around the great points of entry (Santos-São Paulo, Rio, Bahia, Buenos Aires). Moreover, these countries did not have the attraction of a Pacific coast to reach, except in the freezing south of Argentina.

5.2. Moving freely

> ... and if from this land of illiterates, of rustics, of crude artisans, one cannot expect supreme arts and crafts, may they be commissioned from Europe, for my convent of Mafra, paying for them with gold from my mines and other goods, the decorations and ornaments that will make the craftsmen there rich, and we filled with wonder upon seeing the decorations

and ornaments. Portugal requires nothing but stone, tiles and wood to burn and men for brute force, little science. If the architect is German, if the master carpenters and masons and stonemasons are Italian, if English, French, Dutch and other merchants sell and buy from us every day, it is only right that there should come from Rome, Venice, Milan and Genoa, and Liège and France and Holland, the bells and carillons, and the oil-lamps, the lamps, the candlesticks, the bronze torches, and the chalices, the silver ciboriums, the shrines, and the statues of the saints to whom the king is most devoted, and the vestments of the altars, the frontals, the dalmatics, and the chasubles, the copes, the cords, the canopies, the palliums, the white tunics of the supplicants. (José Saramago)[1]

In 1717 Juan V, King of Portugal, decreed that work begin on the construction of the imposing Mafra Convent and the royal palace attached to it, thus fulfilling a vow he had made upon the birth of his heir. Thirteen years and an immense workforce – in the years of greatest effort, amounting to some 40,000 workers – were needed to complete the construction of the enormous complex, which was made possible by the influx of gold from the new mines in Brazil.[2] The workers came mainly from the north of Portugal, but also from Spain and other European lands. We can consider its construction emblematic of the degree of interaction of the Western world in the first part of the eighteenth century: there was gold from Brazil, labour from Portugal and Spain, and technicians and craftsmen from other European countries. Local and international migrations combined in building the Mafra Convent, for varying time periods, indeed in many cases seasonally, because the construction workers were peasants who returned to their villages for work in the fields or harvests. In other cases, the migrant's labour was contracted for a specific job or function, after which he returned home. These movements, however, had the common characteristic of being 'free': they were neither forced nor organized, but mostly the consequence of individual or family choices. We may grant that these choices were sometimes compelled by hardship or hunger, but they still contained an element of voluntary choice. Until the Industrial Revolution, European societies were predominantly peasant societies, tied to the land, but not immobile. There was, as always and everywhere, a local mobility – to go to markets, to the sites of fairs and celebrations, or to farmsteads, villages and towns for small-scale trade. This meant a short-range mobility, with return at the end of the day, or within a few days. There was also longer-distance migration, to cities, which in the eighteenth century began to expand and grow. There was migration towards commercial and maritime hubs, from Seville to London, from Bordeaux to Amsterdam. There was seasonal migration, often from alpine and mountainous regions, travelling long distances mainly in search of work in the fields. Rural populations had deep family or community roots in the land, but this did not prevent the movement of individuals, with a minimum of organization and a fair amount of freedom of movement not similarly visible in the large international migration flows. Of course, there were controls and impediments, often motivated by questions of security and public order; permits, passes and even internal passports to be obtained; occasional taxes to be paid. But on closer inspection, the rural populations were neither immobile nor closed in on themselves. Within certain limits, the seasonal migration was free migration. We have already mentioned Portugal, and the labour mobility required for the construction of the Mafra complex; the situation was no different for the building of the country's other great monasteries and cathedrals, such as the Mosteiro dos Jerónimos in Lisbon, or, in medieval times, the monasteries of Batalha and Alcobaça, themselves engines of internal migration. However, it is worth focusing for a moment on seasonal migration, which was the most frequent and characteristic mode of mobility throughout Western Europe. First, such movements were regular, i.e. annual, and had well-defined periodic cycles, depending on the nature of the locations, the prevalent crops, the land regime, climatic conditions and other characteristics. Seasonal workers were almost always men, living in poor, often mountainous settings and supplementing meagre resources with temporary work away from their own communities, in which there was an overabundance of labour. In the annual migratory cycle, departures and returns depended on the agricultural characteristics of the destination regions; during the sometimes very long journeys on foot to reach them, there might be other opportunities for work or small trade. Sometimes the migration became permanent, perhaps because the men got married in the places where the men had come to work.

In Spain, as in Portugal, seasonal migration moved in a north–south direction, from mountainous areas or those with very fragmentary land structures to the wheat-producing plains of the *meseta* or the Alentejo.[3] In Spain, seasonal workers came from the northern provinces – Galicia, Asturias and Cantabria – for spells of two to three months of work, from late spring to early autumn, for both reaping grain and grape harvesting in the vast territories of the *meseta*. 'The Galician reapers, who left in May–June and returned in early autumn, would make journeys of ten or more days to reach their places of work, where they would be occupied for 50–60 days.'[4] For example, the journey from Lugo, in Galicia, to Madrid, took an eleven-day journey on foot, progressing an average of ten leagues per day.[5] These workers could expect meagre earnings: one example refers to pay being barely sufficient to buy enough bread at the market for a year's survival.[6] Other seasonal migrants, also coming from the northern provinces, worked in construction, as stonecutters, stone masons, bricklayers or plasterers; or they carried out modest craft activities, as boilermakers, rope-makers or cutlers; or even did a little itinerant trade, or transported goods by mule. It is almost impossible to quantify these migration flows, which set tens or hundreds of thousands of people on the move: in the second half of the eighteenth century, for example, it was estimated that 20,000 migrants descended from the Portuguese Trás-os-Montes to the middle and lower Duero valley for grape harvests and other wine-related work. The grain reaping and grape harvesting in Castile and elsewhere attracted tens of thousands of seasonal workers.

With its very different geographical configuration, Italy offers a very varied picture of internal mobility. To stick to seasonal migrations, an early nineteenth-century observer quantified

> seasonal migration from Abruzzo and Campania to the Maremme and Agro Lazio for seasonal work and land reclamation at the end of the eighteenth century. The emigration flows were estimated at 13,000 a year from Abruzzo; while from Campania 7,000 left the Kingdom for the harvest, 1,500 for work in the Roman countryside and 3,000 for work in the Pontine Marshes.[7]

A survey covering 1809–12, during the period of French rule provides a very interesting geographical profile of internal seasonal movements in the Kingdom of Italy (thus excluding the Kingdom of Naples).[8] There were seasonal flows of lumberjacks, woodcutters and coalmen, coming from the Apennine arc and heading to the Maremma coastal strip, almost as far as Rome, or to Corsica, Piedmont or France. Other migratory flows were linked to the construction sector – bricklayers, stone breakers, stonemasons – with a geographical profile similar to that of the woodworkers. Up to 20,000 seasonal workers arrived in the rice fields of Vercelli and Lomellina, the only sector with a major presence of women; another 40,000 arrived in Maremma and the Roman countryside for the harvests. Other migratory flows departed from the Alpine valleys towards the regions on the southern and northern slopes, and the shepherds on the transhumance routes, from the Abruzzi towards the Maremma or the Tavoliere in Puglia, enriching the general picture of mobility.

> In the regions of transhumance and extensive cultivation in central and southern Italy (Tuscan Maremma, Abruzzi, Puglia. . .) . . . seasonal agricultural migration is predominant . . . It takes place over average distances of 100–200 kilometres, and even more. The length of their stay varies according to the type of work and can range from 1–2 months to up to 6 months a year.[9]

Each European region had its own characteristic forms of short-term mobility, as dictated by its climatic, geographical and economic peculiarities. In Italy, as in Spain, France and other European regions, transhumance involved millions of sheep, accompanied by tens of thousands of shepherds, regularly making long journeys at specific times of the year. This mobility was functional to the wool trade and the manufacture of textiles, and in Spain was coordinated by the powerful Council of the Mesta.[10]

From the eighteenth century, accelerating population growth created imbalances between population, territory and resources, an impoverishment of vast layers of Europe's rural communities, and new drives to both temporary and permanent migration. Thus, from France many seasonal migrants moved to Spain; in the harvest season, numerous seasonal migrants flowed from Scotland and Ireland around Britain;[11] in Scandinavia, many thousands of seasonal migrants flocked to the northern coast of Norway each year for fishing activities.[12]

Free mobility did not only concern seasonal migration by agricultural workers, or indeed that connected to certain trades and professions who answered a widespread but hardly intensive demand. In the seventeenth and eighteenth centuries, outright labour markets emerged in Europe that attracted not only permanent immigration, but also a workforce which came for periods of varying length, around important poles of development. This labour market was no longer – or not only – dictated by the rhythm of the seasons. A well-known example is the North Sea labour market built around Europe's most advanced region, the Seven United Provinces (today's Netherlands), and which extended along the entire coastal strip from the Pas-de-Calais to Bremen, Germany.[13] This region was simple to reach and travel across, due to the ease of communication by seas, rivers and the dense network of canals, and it had an advanced economy. Its intensive agriculture, manufacturing and construction industries, its shipyards and maritime activities, its transport and trade attracted some tens of thousands of workers every year.[14] In Germany, it has been observed that a considerable portion of the population in the region between the border and the Netherlands and the cities of Münster, Hannover and Hamburg 'spent every summer in Holland cutting fodder, digging peat for fuel, fishing for herring or making bricks'. This phenomenon intensified in the eighteenth century: 'meanwhile, poor peasants in the Ruhr valley habitually went to the coal mines in the winter, but returned to their fields during the summer', while in the cities there were numerous seasonal or temporary workers, such as carters or bricklayers.[15] At the beginning of the nineteenth century, highly developed labour markets, which attracted a few tens of thousands of workers each year, had formed in the region of eastern England between London and the Humber, and in the Paris region, as well as the predominantly agricultural labour markets of Provence, Languedoc and Catalonia, and the ones in the Italian and Iberian peninsulas, which we have already mentioned. These seasonal or short-term movements were much less intense and widespread in Central and Eastern Europe, perhaps due to the feudal character of land distribution and the state of serfdom in which rural labour was kept, preventing its free movement.

It has been written that 'Each spring the roads of Europe came alive as peasants streamed out of their villages and fell in with the bands from other villages to seek work in distant places, and the roads teemed again in late autumn when the workers began their homeward trek.'[16] This is a romantic summary of a very diverse and unequal phenomenon, driven by free choice in the hardly permissive societies of the time.

5.3. The great transoceanic migration

> O, in eighteen hundred and forty-one
> my corduroy breeches I put on
> to work upon the railway, the railway,
> I'm weary of the railway.
> In eighteen hundred and forty-two
> I did not know what I should do.
> In eighteen hundred and forty-three,
> I sailed across the sea.
> In eighteen hundred and forty-four
> I landed on Columbia's shore.
> In eighteen hundred and forty-five
> When Daniel O'Connell he was alive.
> In eighteen hundred and forty-six
> I changed my trade to carrying bricks.
> In eighteen hundred and forty-seven
> Poor Paddy was thinking of going to heaven
> to work on the railway, the railway,
> I am weary of the railway.[17]

The nineteenth century marks the beginning of a profound discontinuity in transoceanic movements. The colonial model, driven by the interests of the migrants' home countries, dissolved, first with the independence of the United States, then with that of the former Iberian colonies in the 1810s, and with the end of the Napoleonic wars and the continental blockade that had hitherto hindered transatlantic communications. The European continent saw the final disappearance of the mercantilist and populationist ideology according to which emigration meant a loss of wealth for the countries of origin, and was thus a phenomenon to be banned or strictly limited. With the acceleration of demographic growth, the onset of industrialization, and rising agricultural productivity, there were rising numbers of impoverished masses. Emigration appeared an efficient way to get rid of the poor who burdened charitable institutions and city administrations. European states gradually loosened their restrictive or prohibitive legislation against emigration, or removed constraints on it, in a long process that would span the whole nineteenth century. In Great Britain:

> Already before 1830, in fact, to prevent the population of neighbouring Ireland from flowing into its territory for want of a possible American outlet, the British government abolished the restrictions on expatriation. In the years that followed, forms of state support were set up to incentivize migration, while, in the same efforts to promote it, there was also a move to guarantee better travel conditions for emigrants.[18]

There were numerous parliamentary commissions to deal with the issue, and emigration began to be seen not as a means to get rid of a demographic surplus, but as a way to put the emigrants' work to good use, by providing them with cultivable land in the destination countries. In 1840 the Colonial Land and Emigration Department was created; thus began an era of migration that was not only essentially free of constraints, but also enjoyed a measure of support.[19] A similar path was to be followed by the Scandinavian countries, though in their case the laws sanctioning freedom of migration came in the 1860s.[20] In Germany, before the creation of the Northern Confederation in 1867, which adopted a more liberal legislation, a plurality of rules were in force in the various states and principalities, but emigrants who left with a permit from the authorities were usually deprived of citizenship, with the effect that a large share of emigration was clandestine in nature. In Austria, the ordinances of 1784 forbade emigration, both for military reasons and because it was not compatible with the feudal system. Exceptions relied on approval by royal decree. Emigration was, in fact, considered a betrayal of the nation. The 1832 law confirmed the previous one, although it was less harsh and included more exceptions; the emigrant would nonetheless lose his citizenship. The new constitution of 1867, while declaring freedom of migration, made it conditional on military authorization, which was difficult to obtain, and in any case implied the loss of citizenship. Much of the transoceanic emigration that took place was therefore clandestine.[21] In Hungary, the situation was not too different. In Russia, emigration was de facto forbidden and those who did emigrate lost their citizenship: in 1892 changes were introduced to allow the emigration of Jews, under the auspices of a Jewish organization, but this also meant the loss of Russian nationality. In Italy, following unification in 1861, emigration was subject to public security laws; in 1888, then in 1901, organic laws were passed that explicitly authorized freedom of emigration: for work, according to the 1888 law, and then for all forms of emigration in the subsequent legislation (art. 1 of the 1901 law: 'Emigration is free within the limits established by the law in place'). This latter introduced provisions to protect the emigrant and to regulate the anarchy of intermediary agents between migrants and shipping companies.

This brief overview is useful to understanding how, from a legal point of view, the nineteenth century saw a slow turn towards full recognition of emigration as an individual right, accepting the principle of the emigrant's freedom of choice, without burden and with limited preclusions. This is not to say that emigration did not take place in countries and eras which did have restrictive and punitive legislation: it did indeed take place, counting on lenience in the application of the law or else taking advantage of wide holes therein. This increasing freedom to leave Europe was matched by a relative freedom to enter the destination countries. In the United States, European immigration was basically free up until the restrictive laws of the 1920s. It is true that with the increase in arrivals, a plethora of decrees were issued, mainly directed at excluding people considered 'undesirable' because they carried diseases, or were manifestly unable to support themselves, or were considered dangerous to public order. But 'Notwithstanding so formidable an array of restrictive measures . . . it is due to the immigration commissioners to say that for many years they judged cases more on their individual merits than by following the strict letter of the law.' Although the regulations spoke of excluding the poor, 'nobody in authority seemed to know just what constituted a pauper'.[22] Federal legislation was first introduced in 1882, decreeing the exclusion of the Chinese, but confirming freedom of immigration for Europeans, with the exception of the 'undesirable' (and by 1917 this included the illiterate). In Canada, legislation contained elements encouraging immigration, with the usual limitations for undesirables. In Argentina, legislation has always strongly favoured immigration. The same can be said for Brazil and Uruguay, destinations of important migration flows. These countries encouraged immigration with incentives concerning travel, residence and the granting of land.[23]

The – certainly incomplete – statistics on this period give an idea of how much transoceanic emigration grew from the 1840s up until the outbreak of the First World War. From 1846 to 1915, 43 million Europeans emigrated to destinations across the oceans: 18 million from the lands of the so-called 'old emigration' (mainly Britain and Ireland, Germany and Scandinavia) and 25 million from the lands of the so-called 'new emigration' (Mediterranean, Balkan and Eastern European countries), the former prevailing until the 1880s, the latter thereafter and up until the First World War. Approximately two-thirds of this migration flow went to the United States and Canada, the remainder mainly to Argentina, Brazil, Uruguay and Oceania; the maximum outflow – an average of 1.4 million per year – occurred in the decade before the war.

There were various factors behind the great European migration, and they have been the subject of much investigation and research. It was driven by powerful underlying forces, starting with the acceleration of demographic growth on the continent, the cradle of modern demographic transition. Between 1700 and 1800 the European population increased by half (from 121 to 188 million, +55 per cent), but in the following century it more than doubled (404 million in 1900, +115 per cent). The rise in agricultural productivity, combined with population growth, led to land fragmentation, with an increase in the landless or otherwise underemployed and impoverished population. In the second half of the century, competition from the agricultural production taking place on the new continent lowered food prices, prompting further crises in the countryside, and only in the regions with the strongest industrial development was the demand for labour able to absorb the rural unemployed. The powerful globalization process of the second half of the century resulted from the closer relationship between America, rich in natural capital and in need of labour to exploit it, and Europe, which was capital-poor but rich in unused manpower. At the same time, the regulatory constraints placed on mobility were gradually being lowered, while travel by both land and sea became quicker and less costly. The communities forming overseas also helped to make onward travel easier and cheaper. In addition, information about opportunities in the immigration countries, which in previous centuries had relied on proclamations or the action of a limited number of recruitment agents, circulated quickly and reached the furthest corners of the continent. The closer and more frequent relations with the communities of compatriots already settled in the countries of arrival, and the propaganda of the recruiting agents – themselves hired by the large shipping companies – guaranteed knowledge and information that was otherwise inaccessible. 'At one point, a transatlantic shipping company had no less than 3,400 agents in the field who organized the voyages and advanced money from the United States' to British and Irish emigrants. But the companies' agents were seen across Europe as a kind of plague to be dealt with, due to the large number of cheats and swindlers who took advantage of the ignorance and credulity of the migrants. In Italy, the rapporteurs of the 1901 law estimated that there were about 10,000 agents and sub-agents of shipping companies, scattered all over the country, engaged in an unregulated intermediary activity that led to frequent abuses.[24] Finally, the competition among the companies and the logistical and technical progress of shipping caused a fall in both costs and journey times: in the 1850s, a crossing between England and the United States cost $44, but by the 1880s, $20; in the 1890s, a crossing between Spain and South America cost $50, which fell under $35 by 1900.[25]

Here is not the right place to go deeper into the complexities of the great migration – also because, while there were certain common factors, the geographical, economic and political peculiarities of the various regions strongly influenced its development. These pages are, however, dedicated neither to the description nor to the analysis of the great migration, but rather to its distinctive character compared to other migratory movements, both before and after it. It was neither forced, nor organized, nor led by political choices. It is not that these elements were missing entirely: pogroms were a powerful cause of Jewish emigration from Russia; the disaster of the

Great Famine in Ireland drove rivers of people to emigrate; up until the 1830s Australia was populated by the deportation of convicts from Britain; some countries organized settlements in colonized territories. But most of the migration flows of the nineteenth and early-twentieth centuries consisted of individuals and families who left not because they were driven by danger, or by the force of the state, or by active organizations, but rather by their own free, autonomous decision. This autonomy and freedom benefited from the departure countries' recognition of the right to emigrate, combined with the openness of the destination countries. To push the argument somewhat, we can say that in no other period of known history have there been international migratory movements of comparable size and within such a free migration system. In other words, the century of the great migration is almost unique in the history of migration in the West; yet many of the paradigms and theories on migration as a phenomenon have been developed based on this unique example.

In the last century, from the end of the First World War to the present, migration flows have been strongly conditioned and freedom of migration has evaporated. Starting in 1921, destination countries implemented highly restrictive and selective policies, which were then intensified during the Great Crash. The Second World War interrupted all movement, and in the decade 1938–48 mobility took the form of forced migration, deportations or refugee returns. In the forty years that followed, Europe was divided in two by the Iron Curtain; the populations to the East were stripped of their freedom to emigrate; this freedom endured and was strengthened in Western Europe, although it was limited to movement between states now bound by myriad connections. In the last thirty years, since the fall of the Iron Curtain, intra-European migration movements have grown strongly, but they are still migrations within an area populated by European 'citizens'. On the other side of the ocean, the United States maintained many historical restrictions, which were only loosened in 1965, with some subsequent adjustments. South of the Rio Grande, between the countries of Central and South America, there had existed a traditional openness that allowed for travel from state to state without major hindrances; in numerical terms such migration has always been modest. But in the last two decades, even these openings, the heirs to a traditional cultural solidarity, have been closing in.

These brief remarks indicate that the era of free international migration is over – it met its perhaps definitive end as long as a century ago. Free movements still exist, but their perimeter has been restricted to the national space or, at most, to some multinational spheres such as the European Union.[26]

5.4. America: the 'advancing wave' of migration

> Martin was, like most of the inhabitants of Elk Mills before the Slavo-Italian immigration: a Typical Pure-bred Anglo-Saxon American, which means that he was a union of German, French, Scotch, Irish, perhaps a little Spanish, conceivably a little of the strains lumped together as 'Jewish,' and a great deal of English, which is itself a combination of primitive Briton, Celt, Phoenician, Roman, German, Dane, and Swede.[27]

When Sinclair Lewis wrote these lines in the 1920s, the great wave of migration that had brought tens of millions of Europeans from the old to the new continent over the course of a century had come to an end. It had profoundly changed both the ethnic composition and the geography of the US population. In 1790, the white population of the United States was, in its large majority, of British and Irish descent, with a small German and Dutch component.[28] By the early twentieth century, however, all the major European nationalities were represented by sizeable and influential minorities.[29] In 1790, the frontier ran parallel to the Atlantic coast, and not far from it; a century later, it had reached the Pacific, 4,800 kilometres further West. This spread and extension of the settlement of the North American continent was owed to a variety of geographical, economic, social and political factors, too complex to be summarized in a few lines. But this development was moved above all by the free choice of the protagonists themselves. It was a 'free choice' aided by the fact that this migration took place in territories that were either unpopulated or only very sparsely populated, and was able to suffocate or suppress the weak native populations. It was a sort of 'advancing wave' populating territory, across a broad front, through successive waves of migration but was also pushed forward by generations of migrant children.[30] It should also be said as a first consideration that these free choices, which formed the 'advancing wave' we are going to discuss, came to the great detriment of the native populations, the native Americans, who were swept away by this wave. Their stories could rightly be included in the pages we dedicated to forced migration: for example, the consequences of the Indian Removal Act, signed by Andrew Jackson in 1830, on the basis of which tens of thousands of members of the 'Five Civilized Nations' were expelled from their territories in the east and 'removed' west of the Mississippi, a long journey called the Trail of Tears because of the suffering it brought and the victims it left behind.[31] Or, again, in the latter part of the century, the forced confinement of native Americans on reservations west of the Mississippi, based on the Indian Appropriation Act of 1851.

To return to the 'advancing wave', we can better understand its nature by jumping back a few thousand years, to the migration by 'anatomically modern man'. This took place through movement into territories that were either empty or occupied by other human types with less developed abilities, such as the Neanderthals in Europe. We can imagine that these migrations were also free, generated by individual and group choices. The introduction of agriculture in the Near East and Europe is said to have begun 9,000 years ago, during the Neolithic revolution, and reached its conclusion 5,000 years ago in Britain and Ireland. One theory explains the spread of agriculture in this period as a consequence of a spread of culture: it is said that knowledge, practices and techniques travelled, and were adopted by the populations settled in the various territories.[32] According to the alternative 'demic spread' theory, it was instead farmers who migrated, sustained by a more stable and solid demographic growth, and who brought their agricultural production techniques with them. The combination of demographic growth and displacement is thus said to have led to a slow but continuous 'advancing wave' which populated the continent.[33] The dating of the archaeological remains of the sites around Europe where populations developed settled agriculture through cereal crops are consistent with this theory. The peopling of the continent appears to have taken place along a south-east to north-west axis – from the eastern Mediterranean to Britain and Ireland – with a slow penetration of migrants who extended the cultivation of fields into new lands and settled them with houses and villages. A slow 'advancing wave' – favoured by the growth demographic and attracted by the availability of new land – which moved at an average rate of expansion of just over one kilometre per year.[34] This spread-expansion seems not dissimilar to that produced by the Bantu peoples, whose migration from their original spaces, on the border between Cameroon and Nigeria, gradually occupied central and southern Africa. This process took place over the course of three millennia, exporting agriculture over a north–south route spanning almost 5,000 kilometres.[35]

These forms of prehistoric migration – both the potentially faster ones linked to the high mobility of hunters and gatherers, and the slower ones linked to the spread of agriculture – took place in empty territories or, at any rate, ones populated in low numbers and in dispersed fashion. Only rarely did they come into contact with other inhabitants, and did not compete with them for resources. In the history of the last two thousand years, with an increasingly densely settled world, such conditions of uncontested spread have become ever rarer. Migrants have had to contend with local populations, either by negotiating means of coexistence, imposing them or else enduring them, depending on the given power relations and circumstances. Migration processes have generated conflicts, confrontations, mixing and hybridizations – and these have been cultural, social and biodemographic in nature.

In the American case, to which we should now return, we find traces of the Neolithic 'advancing wave' pattern. Before 1800, settlement had begun in the north of the country, from New England to the Ohio valley. Even the territories beyond the Appalachians, as far as the Mississippi valley, were considered outlets for settlers from New England and from the Middle States. 'By 1800 some 300,000 persons were busily growing corn and tobacco in that fertile country' in Kentucky and Tennessee.[36] Population grew in western New York State and Pennsylvania, responding to the supply of land. Ohio had become a state in 1803. In 1812, the presence of 12,000 residents justified the establishment of the state of Illinois, but in Indiana there were already 25,000 settlers and in Ohio 250,000.[37] By the end of the War of 1812, fought against Britain, 'the settled portions of the United States resembled a great triangle, its base along the Atlantic seaboard and its apex at the junction of the Ohio and the Mississippi rivers.'[38] Between 1812 and 1840, the wave of migration and settlement reached the Mississippi; coming from the east, but also fuelled by migrants from Germany and Scandinavia, it extended across the river to Texas. There, the few thousand restless settlers acquired independence from a weak Mexico in 1836, bringing Texas into the Union in 1845. In the same year, California was also taken from Mexico; the discovery of the gold mines led, in 1849 and the following two years, to the arrival of 50,000 prospectors arriving by land and another 25,000 arriving by sea, crossing the Isthmus of Panama or rounding Cape Horn. The many migrants who were left empty handed headed back east to the mountains, continuing their prospecting activities, generating more gold rushes as news of fruitful finds spread. The immigrants who had crossed the Mississippi and populated the territories on the right bank of the river (Dakota, Minnesota, Iowa, Missouri, Arkansas and beyond) faced the boundless expanse of the prairies, all the way to the Rockies: these were huge territories but lacked communication with the more developed areas. Soon the conditions were created for the settlement of these lands. In 1862, Abraham Lincoln had enacted the Homestead Act, which provided for the free allocation of land (strips of 160 acres, or 65 hectares) to those who applied for it; they were granted immediate possession of the land, and the prospect of becoming its permanent owners on condition that they farmed it. The law was revised many times in the following decades, since in regions of low rainfall 160 acres of land proved insufficient for a farm; moreover, countless abuses and swindles were committed in the apportionment of land. However, land was available, partly because the railway companies that had received generous allocations from the government were putting it on the market. In 1869, the intercontinental railway was completed, and the western territories of the country became easily accessible. Public opinion identified with the myth of the Race to the West ('Go West, young man, go West and grow up with your country!'[39]); entrepreneurs, big companies, the government, missionaries (such as the Mormons who colonized Utah) and the press all joined forces in driving the push westward. Hunters, adventurers and explorers made up the vanguards; they were followed by migrants who became ranchers, and then farmers who set up their own farms. In the twenty years between 1870 and 1890, 430 million acres (174,000 square kilometres)[40] were allocated; this was also the same time that the tragedy of the 100,000

Plains Indians unfolded. These latter were subjected to bloody assaults (the war with the Sioux of 1875–6) that forced them to abandon the life of the free hunter for a sedentary one, as they were confined to reservations. The results of the 1890 census made it possible to declare the 'closing of the frontier': there were no more empty territories still to be settled by the migratory wave from the east.[41] The peopling of the North American continent had been completed.

As mentioned above, this was a movement initiated by fur trappers, miners and cattle breeders, but whose stable mass of numbers was made up of pioneer farmers. They were driven by forces analogous to the ones that had sustained waves of migratory advance in past eras: the occupation of empty or half-empty spaces, with prolific farming families, which in turn contributed to later advances. One of the most credited explanations for the larger size and higher birth rate of farmer-proprietor families in the nineteenth century postulates that these developments were the result of the greater availability of land and the lower cost of providing for children.[42] The frontier was advancing not only because of the influx of new migrants from the east – the average worker faced cost barriers to travelling, acquiring land, preparing it for cultivation, buying tools and seeds, and building a house – but also because of the self-propelling force of already settled families.[43] In a profoundly different world, however, the wave was pushed on by immigration from Europe and industrialization that rapidly subverted the old economic order.

The great advancing wave that populated the North American continent cannot easily be reduced to any simple pattern; indeed, as it progressed, the geography of the territories, the composition of the migrants and the available techniques changed. So, too, did government policies, though they were generally favourable to westward expansion. But some concluding considerations do have to be made. The first is that this migration was based on the free choice of the protagonists

– even if with all the limits of the concept of freedom noted in the introduction to this chapter, dedicated to free migration, a rare phenomenon in Western history during the past millennium. But the migrants' freedom meant the loss of freedom for the weak native populations: the survivors of the ill-matched conflicts suffered forced migrations to the areas chosen by the victors, and the overthrow of their centuries-old way of life.

The second important aspect concerns the effects of the great migration on the identity and mentality of the American people. Frederick Jackson Turner was the first of a series of historians who supported the thesis that migration had brought the true 'Americanization' of society, which in its early days had followed a European model of culture, mentality and institutions.[44] This model was no longer useful in the advancement of the frontier, in which individualism, independence, territorial and social mobility and the functionality of simple organizational models were celebrated.

> The cultural baggage lost during migration, the impact of the unique environment, the acculturation that occurred as men of differing racial and geographical backgrounds met and mingled ... all contributed to the uniqueness of the resulting social order. As the new West merged with the East, the pioneers placed their stamp on the civilization of the nation as a whole.[45]

This thesis has been the subject of much criticism and revision in recent decades, but it bears an undoubted kernel of truth. A third aspect consists of the relationship between the oldest settled populations and the 'new' immigrants, i.e. the wave that arrived from Europe from the 1840s. The pioneers and early farmers were natives, accustomed to the use of axes for logging, extensive farming, river navigation, hunting and fishing. In terms of ideas, they paved the way for settlement by migrant farmers, native or immigrant, who often had different cultures and aspirations. Immigrants of Germanic origin tended to create farms, which they then enlarged and integrated with other lands assigned to their children, in order to create close-knit groups of settlers.[46] 'When the German comes, the Yankee goes' was a popular saying.

> But the American father made no such efforts on behalf of his offspring. To be a 'self-made man' was his ideal. He had come as a 'first settler' and had created a farm with his ax; let the boys do the same. One of them perhaps was kept at home as a helper to his aging parents; the rest set out to achieve beyond the mountains or beyond the river what their father had accomplished in the West of his day. Thus mobility was fostered by family policy.[47]

To simplify things in the extreme, American-born whites constituted a pioneering vanguard in the westward expansion, a vanguard which was pushed on and then replaced by British, German and Scandinavian immigrants who populated the territories west of the Mississippi in large numbers, from the 1860s and 1880s, until the 'closing' of the frontier.

Reconsiderations

The histories gathered in this book tell of examples of migration that differ considerably among themselves, whether in terms of their motives, circumstances or consequences. It is difficult to reduce them to a few comparable dimensions, and impossible to interpret them through any one model. The migration brought about by Rome's military activity consisted only of men, the *filles du roi* were exclusively women; the Germans recruited by Catherine the Great were only families; the migration by Goths, Lombards and other barbarian ethnic groups involved entire peoples. The *ápoikoi* Greeks in the Mediterranean numbered only a few thousand; transoceanic migration set tens of millions of individuals on the move. Seasonal migrants in Europe covered short distances, whereas Haitian earthquake and hurricane victims travelled up and down an entire continent. Moving entire ethnic groups from one end of the empire to the other took only a nod from the Inca in Peru, or a decision of the Politburo in the Soviet Union, but millions of individual decisions were needed to settle the North American continent.

Nor can it be assumed, looking at the various forms of migration, that it allows a way out of poverty or at least an improvement in migrants' living conditions. This postulate does apply to the case of free migration, but not to other forms of displacement. It certainly does not apply to forced migration: Africans lived better free in Dahomey than as slaves in the sugar-cane plantations of the Caribbean, and the same is true for almost all groups relocated against their will. Furthermore many instances of organized migration throughout history have also resulted in painful failures, even if they had good initial intentions.

The histories recollected in these pages do not aim to represent an analytical sample of episodes of migration, nor to establish an embryonic 'history' of migration. If anything, they are meant to prompt reflection on certain aspects of migration as a phenomenon. One that I consider absolutely fundamental concerns the degree to which migration can be deemed a 'success'. To measure this, however, a series of conceptual and theoretical – as well as practical – hurdles need to be overcome. Indeed, it must first be asked: 1) what is success?; 2) to whom or to what does it refer?; 3) on what time scale is it to be measured?

The success of migration. There can be various interpretations of what this means, depending on research perspectives and what aspects of migration phenomena we wish to explore. These perspectives can be, for example, biodemographic, anthropological, social or economic. The simplest one is the biodemographic one, and is normally based on measuring certain indices (survival, health, reproductivity) to assess whether migrants and their descendants have a better or worse 'profile' than the populations of origin, in terms of their numerical dynamics. In this respect, the *filles du roi* and their husbands and partners represented a very successful migration, due to their 'founder effect' for a very large descendant population, which contrasted with the stagnant French population of origin. A similar consideration can be made with regard to the German migrants in the Volga region, who increased their population tenfold in less than a century. Conversely, as we have seen, the migration of German settlers to the Maremma had a disastrous outcome. Moving from the micro to the macro level: the African slave trade also had a disastrous outcome; the enslaved communities in the Caribbean or Brazil could only sustain themselves numerically due to the continuous influx of other slaves from Africa, without which they would have vanished.

The biodemographic dimension is not the only criterion of 'success'. It can also be measured in terms of the improvement in the migrant group's living conditions, as compared to variations in the living conditions of the source population. Of course, the construction of an algorithm that gives this definition concrete form is very difficult for the present, and (near-)impossible for the past, due to the complexity of the concept of 'living conditions'. But this concept can be boiled down to some direct or indirect indicator of economic or social well-being. Evidence suggests, for example, that European immigration to the Americas, both North and South, resulted in undoubted progress for migrants and their children's, and often grandchildren's, generations.

Success for whom? It is generally assumed that 'success' relates to migrants, their families and, possibly, their descendants. Migration, however, also affects the 'success' of the populations of origin. Take Irish emigration: if one assumes a Malthusian position (and it makes sense to take one, in this case), emigration 'emptied' the island, improved the ratio of population to natural resources, changed the demographic system, and slowed down its natural dynamics (with its late marriage and high rates of celibacy and unmarried women). In simple biodemographic terms, it was a disaster; in terms of living conditions, it was a success. In the case of the short-lived migratory movements in the European labour markets in the seventeenth and eighteenth centuries, they were a success both for the departure populations, who benefited from the income contributions of their migrant compatriots, and for the destination populations, who needed labour and external contributions.

The time scale. The longer the time scale, the more diluted and complex the judgement on the 'success' of migration becomes, if only because of the possible intervention of other factors, or even because the migrant group loses cohesion due to mobility or reproductive mingling or other reasons. We can take the example of European emigration to America: a definite success in terms of the well-being of the migrants, their children and grandchildren. But afterwards? North America continued to grow, creating new prosperity; South America went into crisis and this also affected the well-being of the migrants' grandchildren and great-grandchildren. In the 1930s, Italians in Argentina were more prosperous than their compatriots who remained at home, but by the 1960s this situation was reversed, as a consequence of Italian economic growth and the deep crisis in Argentina. Judging by the splendour of Magna Graecia and, three centuries after its foundation, the defeat of the Athenians in the war with Syracuse, we might be tempted to say that the emigration of the *ápoikoi* was a good idea, crowned with success. But what is the point of a comparison with the situation three centuries later? It seems reasonable and realistic not to extend the comparison beyond the span of two or three generations – migrants and their children, perhaps even grandchildren – that form a vertically extended family group, over not-too-long an interval of time.

The study of migration – our fifteen stories suggest – cannot do without an in-depth examination of two closely linked aspects. These are the 'selection' of migrants and their 'fitness'.[1] Migrants are never a representative sample of the source population, and they will often exhibit particular biophysical characteristics and abilities. A 'selection' takes place in terms of one or more characteristics, whether by way of individual choices (the sick and weak do not undertake long and dangerous journeys), because organizers make such a choice (the peasants chosen by the populators in the *Drang nach Osten* were probably more skilled or strong than average; the *filles du roi* had to be 'hardy' and apparently good breeders, etc.), or for other reasons. Selection also influences the 'fitness' of migrants, i.e. their suitability, or ability to benefit from migration, by means of the work and activities they are able to perform in the destination countries, and their ability to insert themselves profitably into a different society. Selection and fitness are thus essential aspects for interpreting the consequences of a migratory movement.

However, this selection is often temporary in nature, inherent to the migrants' generation and not applicable also to their descendants: this is true, for example, of the migrants' better survival rates – for the reasons given above – or of the migrants' aptitudes and human capital, which can only partly be passed on to their children. There is a certain regularity to the fact that the most gifted people, in terms of ability or resourcefulness, are more frequently present in migratory flows: family strategies organized the migration of the relative with the best chance of 'making it'; migration organizers tried to select those most suitable to carry out the expected tasks. In contemporary times, it is the arrival countries that mount other forms of selection. In the nineteenth century, the destination countries for transoceanic immigration excluded those who had physical deformities, or who were suspected of carrying contagious diseases. In our own century, selection mechanisms have proliferated, consistent with very restrictive migration policies that limit the possibility of immigration to very specific categories. In the first phase of the great transoceanic migration, destined to gradually settle the North American continent, families that produced children capable of pushing the advancing wave towards the Pacific played a privileged role. But in the later phase of immigration to urban areas, or of migrants employed in non-agricultural activities, the highly mobile and adaptable male migrant, at first without the burden of a family, was the more likely candidate.

I will end with a general remark. Moving and changing one's abode is a prerogative of human beings. This prerogative may be used or left unused, but it nonetheless remains present. In the history of the West – and of the world – this prerogative has been used with the most diverse motives, and in various forms and circumstances; its consequences have ranged from total success to absolute disaster. The model of free migration, the result of conscious decisions, evaluations and choices, is surely the most preferable one. It is a model that prevails within Western societies, where it remains indisputable, yet it only applies to migration within national borders, or within enlarged spaces such as the European Union. In the case of international migration, it is a century since any such model has existed. Migration is not driven by free individual decisions, but conditioned by a tangle of rules that decide who is entitled to migrate, regardless of individual inclinations.[2] Many of these rules are misguided, do harm to migrants and prevent migration from realizing its beneficial effects. Straightening out these rules is a difficult task and will require extensive efforts from the international community.

Notes

I. Antiquity

1 Seneca, *Of Consolation: To Helvia*, translated by Aubrey Stewart, in *Minor Dialogues: Together with the Dialogue on Clemency*. London: George Bell, 1889, VI, pp. 326–7. I became familiarized with this beautiful text thanks to my friend and colleague Elio Lo Cascio, who pointed it out to me many years ago.
2 Ibid., VII, pp. 327–8.
3 Ibid., p. 328.
4 Ibid.
5 Thucydides, *The Peloponnesian War*, translated by Martin Hammond. Oxford: Oxford University Press, 2009, p. 309.
6 Ibid.
7 Ibid.
8 Ibid., pp. 309–10.
9 Cicero, *On the Republic. On the Laws*, translated by Clinton W. Keyes. Cambridge, MA: Harvard University Press, 1928, p. 119.
10 Franco De Angelis, 'Estimating the Agricultural Base of Greek Sicily', in *Papers of the British School at Rome*, 68, 2000, pp. 111–48.
11 Ian Morris, 'The Growth of Greek Cities in the First Millennium BC', Princeton/Stanford Working Papers in Classics, 120509, 2005, p. 3.

12 *Res Gestae Divi Augusti*, translated by F.W. Shipley, 3. Cambridge, MA: Harvard University Press, 1924, p. 349.
13 Walter Scheidel, 'Roman Population Size. The Logic of the Debate', Princeton/Stanford Working Papers in Classics, 070706, 2007, with a discussion of the different positions on p. 6.
14 *Res Gestae Divi Augusti*, 28, p. 393.
15 Michael Rostovtzeff, *The Social and Economic History of the Roman Empire*, Vol. 1. Oxford: Clarendon Press, 1967, p. 33.
16 Mario Attilio Levi, *L'Italia antica. I: Dalla preistoria all'unificazione della penisola (42 a.c.)*. Milan: Mondadori, 1968.
17 Greg Woolf, 'Moving People in the Early Roman Empire', in Elio Lo Cascio and Laurens E. Tacoma (eds), *The Impact of Mobility and Migration in the Roman Empire*. Brill: Leiden, 2016.
18 Alessandro Barbero, *Barbari. Immigrati, profughi, deportati nell'Impero Romano*. Rome: Laterza, 2006, p. v.
19 Tacitus, *On Germany*, translated by Thomas Gordon. Text from sourcebooks.fordham.edu.
20 Suetonius, *Lives of the Twelve Caesars*, Tiberius, 9. Cambridge, MA: Harvard University Press, 1914, p. 325.
21 Barbero, *Barbari*, p. 15.
22 Paul the Deacon, *History of the Lombards*, II, 6, translated by William Dudley Foulke. Philadelphia: Pennsylvania University Press, 1907, p. 61.
23 Barbero, *Barbari*, p. 15.
24 This translation from *In Praise of Later Roman Emperors: The Panegyrici Latini*. Berkeley, CA: University of California Press, 1994, pp. 121–2.
25 Zosimus, *New History*. London: Green and Chaplin, 1814, p. 104, translation lightly edited.
26 Divided into Grutungi (later Ostrogoths), or Goths of the east, and Tervingi (later Visigoths), or Goths of the west.
27 Ammian, *The History*, translated by J.C. Rolfe, book XXXI, available at https://penelope.uchicago.edu/Thayer/E/Roman/Texts/Ammian/31*.html.
28 Peter J. Heather, *The Visigoths from the Migration Period to the*

Seventh Century. An Ethnographic Perspective. Woodbridge: Boydell & Brewer, 1999, p. 55.
29 Zosimus, *New History*, Book II, chapters 7–8.

II. In the Hands of the State

1 Alexander von Humboldt, *Political Essay on the Island of Cuba: A Critical Edition.* Chicago, IL: University of Chicago Press, 2011, p. 144.
2 See the SlaveVoyages website with the associated database: Estimates (slavevoyages.org). See also Philip D. Curtin's seminal work *The Atlantic Slave Trade. A Census.* Madison, WI: University of Wisconsin Press, 1969.
3 https://en.wikipedia.org/wiki/Pe%C3%A7a
4 Inca Garcilaso de la Vega, *Comentarios Reales.* Caracas: Fundación Biblioteca Ayacucho, 1985, Vol. 1, III/19, p. 161. Garcilaso 'el Inca' de la Vega, 1539–1616, was son of the conquistador Sebastián de la Vega and the Inca princess Chimpu Ocllo. A descendant of Huayna Cápac, having been born in Cuzco, he headed back to Spain in 1560. At first a soldier, then a scholar and writer, in 1609 he published his famous *Royal Commentaries*.
5 Pedro Cieza de León, 1512–54. At the age of twenty he left Seville for the Americas, travelling first in what is today Colombia, then in the vast Inca Empire, spanning its length and breadth, until his return to Spain in 1554. An eyewitness to the civil wars in Peru, he was with Pedro de la Gasca, the crown's representative in Peru, from whom he received the official position of *cronista de Indias*. He wrote a monumental *Crónica del Perú*, an authoritative source on the country's affairs.
6 Pedro Cieza de León, *Segunda parte de la Crónica del Perú*, edited by Marcos Jiménez de la Espada. Madrid: Manuel Gines Hernández, 1880, chapter 22, pp. 84–5, 88–9.
7 Inca Garcilaso de la Vega, *Comentarios Reales.* Caracas: Fundación Biblioteca Ayacucho, 1985, Vol. 2, VII/1, p. 85.
8 Massimo Livi-Bacci, *Conquest: The Destruction of the American Indios.* Cambridge: Polity, 2008, p. 170.

9 Francisco Álvarez de Toledo was Viceroy of Peru from 1569 to 1581. He governed the colony with harsh efficiency, reforming its institutions with his *Ordenanzas* issued after a long and thorough *Visita*, breaking down the resistance of the last Inca Túpac Amaru, imposing regulation on mining activities and overseeing the redistribution of the population.
10 Livi-Bacci, *Conquest*, p. 177.
11 Ibid., pp. 177–8.
12 Luis Capoche, *Relación general de la Villa Imperial de Potosí*. Madrid: Atlas, 1959, p. 77.
13 Franz Werfel, *The Forty Days of Musa Dagh*, New York: Random House, 1934, p. 428; letter from the besieged Armenians, intended for delivery to the US consul-general Jackson in Aleppo.
14 Servet Mutlu, 'Late Ottoman Population and its Ethnic Distribution', *Turkish Journal of Population Studies*, 25, 2003, p. 18. According to Mutlu, the total population in 1914 was 18.806 million, including 1.75 million Greeks, 1.596 million Armenians and 286,000 from other minorities. The estimates are based on the official count from 1914, and a revision of the numbers which appear in the census of 1906–7, which Mutlu corrects in order to take into account the very high under-registration, especially in Anatolia's eastern provinces.
15 Marcello Flores, *Il genocidio degli armeni*. Bologna: il Mulino, 2006, p. 63.
16 Vahakn N. Dadrian, *The History of the Armenian Genocide*. Oxford: Berghahn Books, 1997, pp. 218–26.
17 Ronald G. Suny, 'Armenian Genocide', in *International Encyclopedia of the First World War*, 2015, https:// encyclopedia .1914-1918-online.net/article/armenian_genocide.
18 Franz Werfel, *The Forty Days of Musa Dagh*. Jaffrey, NH: Verba Mundi, 2012, pp. 100–1.
19 Erik-Jan Zürcher, 'The Late Ottoman Empire as Laboratory of Demographic Engineering', *Il mestiere di storico*, 1, 2009, p. 9; there were 300,000 according to Hans-Lukas Kieser, 'Minorities

(Ottoman Empire/Middle East)', in *International Encyclopedia of the First World War*, p. 5, https://encyclopedia.1914-1918-online.net/article/minorities_ottoman_empiremiddle_east.

20 Zürcher, 'The Late Ottoman Empire', p. 11.
21 So named after 1906–11 conservative prime minister Pyotr Arkadyevich Stolypin, author of policies which harshly repressed dissent, as well as extensive land reform and support for the Trans-Siberian Railway.
22 Aleksandr Solzhenitsyn, *The Gulag Archipelago, 1918–1956, An Experiment in Literary Investigation*. London: Vintage, 2018, p. 129.
23 Ibid., pp. 129–30.
24 The deportation of the kulaks, or rich peasants. See note 26.
25 Frank Lorimer, *The Population of the Soviet Union. History and Prospects*. Geneva: League of Nations, 1946, pp. 172–4.
26 Ibid., p. 163.
27 There were three waves of dekulakization in 1930–2; according to Nicolas Werth, the number of people deported to the Urals, Western Siberia, the Far North and Kazakhstan in 1930–1 amounted to over 1.8 million (Nicolas Werth, *Mass Crimes under Stalin, 1930–1953*, 14 March 2008, https://www.sciencespo.fr/mass-violence-war-massacre-resistance/en/document/mass-crimes-under-stalin-1930-1953.html); according to Pohl, the total number of kulaks deported to remote districts in the decade between 1930 and 1940 was close to four million (J. Otto Pohl, *The Deportation and Destruction of the German Minority in the USSR*, 2001, norkarussia.info). For Molotov, who was responsible for the entire dekulakization process, it was 1.3–1.5 million families, with six million members (Massimo Livi-Bacci, *I traumi d'Europa*. Bologna: il Mulino, 2020, p. 87).
28 Olga Chudinovskikh and Mikhail Denisenko, 'Russia; A Migration System with Soviet Roots', Migration Policy Institute, 2017: https://www.migrationpolicy.org/article/russia-migration-system-soviet-roots.
29 Gijs Kessler, 'The Passport System and State Control over

Population Flows in the Soviet Union, 1932–1940', *Cahiers du monde russe*, 42(2–4), 2001, 477–503.
30 Lorimer, *The Population*, pp. 172–3.
31 Rebecca Manley, 'The Perils of Displacement: The Soviet Evacuee between Refugee and Deportee', *Contemporary European History*, 16(4), 2007, p. 496.
32 Ibid., pp. 496, 499.
33 Ingria is a region lying on the southern shore of the Gulf of Finland.
34 For more details, see Werth, *Mass Crimes*. See also the Wikipedia article 'Population Transfer in the Soviet Union', https://en.wikipedia.org/wiki/Population_transfer_in_the_Soviet_Union, with descriptions of the various episodes of forced migration.
35 Generic indications without precise references to sources and to the timespan concerned: the mortality rate for Chechens and Ingush would seem to have been around 20%; for the Kalmyks 17%; for the Mesketi Turks 15%; for Koreans 23%; for Balkans 20% (see 'Population Transfer').
36 Pohl, *The Deportation*.
37 Joseph B. Schechtman, *European Population Transfers, 1939–45*. Oxford: Oxford University Press, 1946, p. 383.
38 Ibid. p. 385.
39 Pohl, *The Deportation*.

III. Misdeeds of Nature

1 Euclides da Cunha, *Os sertões*. Porto: Lello & Irmão Editores, 1983, vol. I, pp. 170–1. The *sertão* is a large semi-arid sub-equatorial region in north-eastern Brazil, with low, irregular and often violent rainfall, and frequent episodes of drought characterized by *caatinga*, large areas with low, winding shrub vegetation. The *retirantes* are the people displaced by drought.
2 A 1946 law defined the *polígono das secas*, a 1.1 million square kilometre region comprising eight states in the north-east of the country.
3 María Verónica Secreto, 'A seca de 1877–1879 no Império do

Brasil: dos ensinamentos do senador Pompeu aos de André Rebouças: trabalhadores e mercado', *História, Ciências, Saúde*, 27(1), 2020, pp. 33–51; Pedro Henrique Barreto, 'História – Seca, fenômeno secular na vida dos nordestinos', in *Desafios do desenvolvimento*, 6(48), 2009, 10 March 2009, available at https://www.ipea.gov.br/desafios/index.php?option=com_content&view=article&id=1214:reportagens-materias&Itemid=39; Edson Holanda Lima Barboza, 'Retirantes cearenses na província do Amazonas, 1878–79', *Revista Brasileira de História*, 35 (70), 2015, pp. 131–55.

4 The project was resumed in 1907 and the railway was opened in 1912, but lasted only a few years, before the rubber industry collapsed due to competition from Asian plantations. In 1972 the railway, the construction of which had cost the lives of 1,600 workers who died from accidents and diseases, mainly malaria, was decommissioned. Built in agreement with Bolivia (which ceded the vast territory of the present state of Acre to Brazil in 1904), the railway was meant to foster the development of this remote Amazon region on the basis of the rubber industry.

5 Letícia Lustosa Martins, 'Varíola em Fortaleza: marcas profundas de uma experiencia dolorosa (1877–1881)', dissertation, Universidade Estadual do Ceará, 2012, p. 130, https://siduece.uece.br/siduece/trabalhoAcademicoPublico.jsf?id=72977.

6 The Great Plains region comprises eight states in the USA (North Dakota, South Dakota, Nebraska, Colorado, Oklahoma, Kansas, Arkansas and Texas) and three Canadian provinces (Alberta, Saskatchewan and Manitoba).

7 Land appropriations legalized by the Homestead Act of 1862.

8 Robert A. McLeman et al., 'What We Learned from the Dust Bowl: Lessons in Science, Policy and Adaptation', *Population and Environment*, 35, 2014, p. 419.

9 Ibid., p. 429. Donald Worster, *Dust Bowl: The Southern Plains in the 1930s*. Oxford: Oxford University Press, 1979; Jason Long and Henry E. Siu, 'Refugees from Dust and Shrinking Land: Tracking the Dust Bowl Migrants', NBER Working Paper 22108, 2016.

10 For the demographers reading this: the large excess of births over deaths was far less than the negative balance between immigrants and emigrants.
11 The best known was *The Grapes of Wrath*, the novel published by John Steinbeck (in 1962 Nobel laureate for literature) in 1939, and adapted the following year for the big screen by John Ford, with Henry Fonda as the lead.
12 James N. Gregory, *America Exodus: The Dust Bowl Migration and Okie Culture in California*. Oxford: Oxford University Press, 1989.
13 Walter J. Stein, *California and the Dust Bowl Migration*. Westport, CT: Greenwood Press, 1973.
14 Gonzalo Fernández de Oviedo, *Historia general y natural de las Indias*. Madrid: Atlas, vol. I, 1992. Columbus, with four ships, foreseeing the hurricane, had taken refuge in a bay not far from Santo Domingo and warned the fleet to delay their departure. Columbus had landed on the island at the beginning of his fourth voyage exploring the region. The story is also told by Bartolomé de Las Casas, who had also seen the extraordinary nugget, weighing about 16 kilogrammes. Among those missing were Commendator Bobadilla, governor of the island, called back home by the sovereigns, and Antonio de Torres, commander of the fleet. The fleet also carried the crown's share of the gold from the mines.
15 Hurricane Jeanne, 2004, 1,300 dead and 1,056 missing; hurricanes Gustave, Fay Hanna and Ike, with 793 dead and 310 missing in 2008; Matthew with 546 dead and 128 missing in 2016. But the island is hit by destructive hurricanes and tropical storms every year.
16 In 2015–20, life expectancy at birth was 63.5 years, compared to 72.5 in the Caribbean region and 74.9 in Central America.
17 United Nations, *International Migrant Stock 2020*, available at https://www.un.org/development/desa/pd/content/international-migrant-stock.
18 Reginald DesRoches et al., 'Overview of the 2010 Haiti Earthquake', *Earthquake Spectra*, 27(S1), October 2011, pp. 1–21.

19 MINUSTAH, an acronym of the French 'Mission des Nations Unies pour la Stabilisation en Haïti', was created in 2004 after the ousting of president Jean-Bertrand Aristide, in order to kickstart a democratization process.
20 Renata Bessi, 'Haitianos cambian rutas de migración tras crisis en Brasil', 9 March 2018, https://subversiones.org/archivos/131801.
21 Cedric Audebert, 'The Recent Geodynamics of Haitian Migration in the Americas. Refugees or Economic Migrants?', *Revista Brasileira de Estudos de População*, 34(1), 2017, pp. 55–71.
22 Massimo Livi-Bacci, 'A Caribbean Odyssey', *Neodemos*, 15 June 2018, https://www.neodemos.info/2018/06/15/una-odyssey-caribbean/.
23 Diego Pons, 'Climate Extremes, Food Insecurity, and Migration in Central America: A Complicated Nexus', Migration Policy Institute, 18 February 2021.
24 The 'dry corridor' (or drought corridor) includes parts of Nicaragua, Honduras, El Salvador and Guatemala. Since 2014 it has been hit by an unprecedented drought cycle that has brought heavy losses of produce and harvests, the migration of farmers and farm labour, and precarious food conditions; it is considered, alongside violence and generalized poverty, to be one of the factors behind the migratory push northwards. This corridor encompasses 4.1 million inhabitants in Honduras, 1.1 million in Guatemala and 0.5 million in El Salvador.
25 'Beast' is the generic name used to define any rail freight transport.
26 These are two excerpts from John F. Kennedy's speeches during his official visit to Ireland in June 1963: the first, in the presence of President Éamon de Valera, on arrival at Dublin Airport, 26 June 1963; the second in Cork City Hall, 28 June 1963. Text online at https://www.presidency.ucsb.edu/.
27 The estimate for 1790 was made by the Bureau of the Census on the basis of family surnames, and does not include immigrants from Ulster, called Scotch-Irish, mostly Presbyterians. See Maurice R. Davie, *World Immigration*. New York: Macmillan,

1936, pp. 41–4. The 2019 estimate can be found in the *American Community Survey*.

28 Downy mildew (*Phytophthora infestans*) is a microorganism (class of Oomycetes), of American origin, which spread in Ireland and Europe at the beginning of the nineteenth century, and which particularly affects Solanaceae such as potatoes and tomatoes. Downy mildew affects the foliage and then the tubers, which rot and become inedible; it is particularly aggressive in wet and cool seasons.

29 There is a boundless bibliography on the Great Famine. Excellent references include Robert D. Edwards and Thomas D. Williams (eds), *The Great Famine*. New York: New York University Press, 1957; Cormac Ó Gráda, *Black '47 and Beyond: The Great Irish Famine in History, Economy and Memory*. Princeton, NJ: Princeton University Press, 1999; and Joel Mokyr and Cormac Ó Gráda, 'Famine, Disease and Famine Mortality: Lessons from the Irish Experience', University College Dublin, Center for Economic Research, Working Paper, 12, 1999.

30 Connell estimates that the population of the island was 2.791 million in 1712 and 3.191 million in 1754. Kenneth H. Connell, *The Population of Ireland, 1750–1845*. Oxford: Clarendon Press, 1950.

31 Connell, *The Population of Ireland*, pp. 81–2.

32 Ibid., pp. 90–1.

33 Ibid., p. 133.

34 There is a wealth of documentation on this topic, summarized in Massimo Livi-Bacci, *A Concise History of World Population*, 6th edn. Chichester: Wiley Blackwell, 2017, pp. 63–4.

35 William F. Adams, *Ireland and Irish Emigration to the New World: From 1815 to the Famine*. New York: Russell & Russell, 1967, pp. 413–14; for migration statistics for the first half of the nineteenth century, see the appendix to Adams's study.

36 Ibid., pp. 238–9.

37 Joel Mokyr, *Why Ireland Starved: A Quantitative and Analytical History of the Irish Economy, 1800–1850*. London: Allen & Unwin, 1983.

38 To grasp the size of the exodus, a comparison with Italy is useful. In the peak decade of Italian transoceanic emigration (1904–13) 3.5 million migrants left; in the peak decade of Irish emigration (1845–54) 1.8 million left. Italy had 34.7 million inhabitants in 1901, so the loss amounted to about 10 per cent of the population; Ireland had 6.5 million inhabitants in 1851, meaning that the loss owing to emigration, at 28 per cent, was almost three times as much.

39 Stanley C. Johnson, *A History of Emigration from the United Kingdom to North America, 1763–1912*. London: Frank Cass, 1913 (Appendix I for migration statistics); D.A.E. Harkness, *Irish Emigration*, in Walter F. Willcox (ed.), *International Migrations*. II: *Interpretations*. Cambridge, MA: NBER, 1931, http://www.nber.org/books/will31-1.

40 Johnson, *A History*, pp. 101–9.

41 Livi-Bacci, *Concise History*, p. 94.

IV. Organized Migration

1 Letter from Colbert to the Bishop of Rouen, M. de Harlay, 1670, reproduced in Thomas Chapais, *Jean Talon, Intendant de la Nouvelle France*. Quebec, Demers, 1904, chapter XVI.

2 While it can be said to have amounted to a few tens of thousands, estimates vary widely. French settlement covered the lower reaches of the Saint Lawrence valley from Montreal to Quebec, in a narrow area of about 35,000 square kilometres. The present-day province of Quebec, which stretches from the US border to the Great North in the eastern part of Hudson Bay, is more than 1.5 million square kilometres in size.

3 Gemery has estimated that New England had 51,500 inhabitants in 1670. Henry H. Gemery, 'The White Population of the Colonial United States, 1607–1790', in Michael R. Haines and Richard H. Steckel (eds.), *A Population History of North America*. Cambridge: Cambridge University Press, 2000, p. 150.

4 'Contracted workers', who were compulsorily bound to an employer for the duration of a few years, in exchange for a pre-

determined fee. At the end they could return to France or settle permanently in Quebec.
5 Gemery, 'The White Population', p. 150.
6 The data relating to Quebec are mostly taken from Hubert Charbonneau et al., *Naissance d'une population. Les Français établis au Canada au XVIIe siècle*. Paris: Ined, 1987. See also Hubert Charbonneau et al., 'The Population of the Saint Lawrence Valley, 1608–1760', in Haines and Steckel, *A Population History*, pp. 99–142; Hubert Charbonneau, *Vie et mort de nos ancêtres*. Montréal: Presses de l'Université de Montréal, 1975.
7 Joseph Schumpeter, *History of Economic Analysis*. London: Routledge, 2006, p. 251.
8 Jean Talon had been appointed as intendant of Nouvelle France in 1665; the crown had recently taken over the administrative prerogatives hitherto devolved to the shipping companies. He held the position until his final return to France in 1673. See Chapais, *Jean Talon*.
9 Charbonneau et al., *Naissance*, p. 6. These are relatively generous figures, given that, at the same time, the average annual salary of domestic servants was of the order of 90 *livres tournois* (notarial deeds of contract). See Arnaud Bessière, 'Le salaire des domestiques au Canada au XVIIe siècle', *Histoire, Économie & Société*, 27(4), 2008, p. 40.
10 Chapais, *Jean Talon*, chapter XIV.
11 On the *filles du roi*, see Yves Landry, *Les filles du roi au XVIIe siècle*. Ottawa: Leméac, 1992. The data included in the following paragraph are taken from this very detailed study.
12 Charbonneau et al., *Naissance*, pp. 9–10.
13 7–10 per cent: see ibid., p. 111.
14 Ibid., pp. 116, 123.
15 Quoted by Martyn Rady, 'The German Settlement in Central and Eastern Europe during the High Middle Ages', in Nora Berend (ed.), *The Expansion of Central Europe in the Middle Ages*. Farnham: Ashgate, 2012, p. 192. The collection edited by Berend contains essays published in the last century, of paramount

importance for understanding the eastward Germanic expansion. Helmold of Bosau was the author of the *Chronica Slavorum*, written in Latin in the second half of the twelfth century.

16 *Drang nach Osten*, or drive to the East, is an expression coined in the nineteenth century to describe the Germanic migration and the Germanization of the Slavic and Baltic peoples, which was later favoured by National Socialism to justify its aggression against the peoples of Eastern Europe.

17 The territories between the rivers Elbe and Oder were settled by Slav (or Vend) populations who occupied the lands abandoned by the Germans who migrated southwards from the sixth century onwards. The Slavic populations across this vast territory then became differentiated, assimilating the Germanic elements that remained there and taking on local names and identities. We refer here, generically, to Slavs (or Vends) meaning the populations settled in the territories which became destinations for Teutonic (Flemish, Dutch) immigration from the eleventh century onwards.

18 I borrow these concepts from Luigi L. Cavalli-Sforza, who described the spread of agriculture in Europe from 9,000 years ago until around 3,000 BC as owing to successive generations of migrants who 'brought along' the new knowledge and techniques of cultivation, starting from the south-eastern Mediterranean and moving north-westwards towards Britain and Ireland. See Albert J. Ammerman and Luigi L. Cavalli-Sforza, *La transizione neolitica e la genetica di popolazioni in Europa*. Turin: Bollati Boringhieri, 1986.

19 'Advancing wave', i.e. a self-propelled migratory process, in the sense that the generations of migrants who settle in a given territory generate descendants who in turn migrate and settle further along in the same direction. This mechanism is repeated across subsequent generations. Thus a migratory movement proceeds in stages and segments, without needing to be continually replenished by the original population. See pp. 122–3.

20 Walter Kuhn, *Geschichte der deutschen Ostsiedlung in der Neuzeit*, 2 vols. Cologne: Graz, 1955.

21 Hermann Aubin, 'German Colonisation Eastward', in *The Cambridge Economic History of Europe*, vol. 1, 2nd edn, Cambridge: Cambridge University Press, 1966.
22 This Germanic population was formed due to the combination of relatively large numbers of immigrants, presumably high reproduction rates, and also – perhaps above all – the assimilation and Germanization of Slavic and Baltic populations.
23 Clifford T. Smith, *An Historical Geography of Western Europe Before 1800*. London: Longman, 1978, p. 176.
24 Rady, *The German Settlement*, pp. 32–5.
25 Aubin, 'German Colonization Eastward', p. 455.
26 'Proclaimed at Peterhof on July 22, 1763, the second year of my reign.' A similar invitation, but drafted in vaguer terms had been proclaimed in December 1762, with little success. Text from Fred C. Koch, *The Volga Germans. In Russia and in the Americas, from 1763 to the Present*, University Park, PA: Pennsylvania State University Press, 1977, p. 13.
27 The 1790 census counted 277,000 inhabitants of German origin, or 8.7 per cent of the white population of the United States. See Hans Fenske, 'International Migration: Germany in the Eighteenth Century', *Central European History*, 13(4), 1980, p. 344.
28 The Treaty of Hubertusburg of 15 February 1763.
29 Roger P. Bartlett, *Human Capital. The Settlement of Foreigners in Russia 1762–1804*. Cambridge: Cambridge University Press, 1979, pp. 58–9.
30 Koch, *The Volga Germans*, pp. 7–8. For a comprehensive history of the Volga Germans, see also Igor R. Pleve, *The German Colonies on the Volga: The Second Half of the Eighteenth Century*. Lincoln, NE: American Historical Society of Germans from Russia, 2001. The website https://www.volgagermans.org/history/migration-russia is rich in information on the history of the Volga Germans, with an extensive bibliography. Much of the information on these pages is taken from these sources. The best historical work is perhaps Bartlett's *Human Capital*.
31 Koch, *The Volga Germans*, p. 19.

32 Catherine's summer residence was located in Oranienbaum, and events were organized there to pay tribute to the empress.
33 Koch, *The Volga Germans*, p. 21. There were many deaths and escapes during the journey; official statistics report that out of the 26,509 migrants who left St. Petersburg, 3,293 did not arrive in Saratov; see Bartlett, *Human Capital*, p. 9.
34 This was a truly vast area on both banks of the Volga. To the east (left bank) it was a flat strip of land many kilometres deep, extending 240 kilometres along the river to both the north and south of Saratov. To the west, along the right bank of the Volga, the land was hilly, and extended south of Saratov for about 120 kilometres, in a belt between 80 and 100 kilometres deep, an area equivalent to a medium-sized Italian region. For this and more, see Koch, *The Volga Germans*, pp. 22–9.
35 Ibid., p. 21. In 1764, a rouble contained 18 grams of silver, and in St. Petersburg a kilo of meat was worth between 4 and 9 kopecks (rouble cents) on the market. See Boris N. Mironov, 'Wages and Prices in Imperial Russia, 1703–1913', *The Russian Review*, 69(1), 2010, p. 51.
36 Koch, *The Volga Germans*, pp. 35–6.
37 Isabel de Madariaga, *Russia in the Age of Catherine the Great*. New Haven, CT: Yale University Press, 1981, p. 362. Pugachev was the leader of the Cossack rebellion of 1773–4 which devastated the lower Volga region. He was executed in 1775.
38 The imperial census in 1897 was the first to follow modern criteria of accuracy. A total of 166,528 Germans were counted in Saratov and 224,336 in Samara; these were individuals whose mother tongue was German. The total for the whole empire was 1,790,000.
39 See https://www.volgagermans.org/history/migration-russia.
40 William H. McNeill, *Europe's Steppe Frontier*, Chicago, IL: University of Chicago Press, 1964, pp. 181ff. Bartlett, *Human Capital*, describes the 'settlement plan' for New Russia on pp. 108ff., and the characteristics and origins of the immigrants in his chapter 4.

41 De Madariaga, *Russia*, p. 365.
42 Lorimer, *The Population*, p. 10.
43 For a balance sheet of migratory movements in eighteenth-century Germany, see Fenske, *International Migration*, pp. 343–7. Prussia made up the losses owing to war and Silesian colonization with an immigration estimated at 300,000 people (p. 346); Hungary, again according to Fenske, absorbed 350,000 German immigrants.
44 Giuseppe Parenti, 'Tentativi di colonizzazione della Maremma nel XVI–XVIII sec.', *Economia*, XVI(1–2), 1937, 43–60. Lorenzo Del Panta, 'La vicenda delle colonie lorenesi in Maremma (XVIII secolo) come esempio di studio di demografia differenziale', in Anna Grassi (ed.), *Statistica e demografia: un ricordo di Enzo Lombardo tra scienza e cultura*. Rome: Università di Roma La Sapienza, Facoltà di Economia-TIPAR, 2007, pp. 141–53.
45 Cayetano Alcázar Molina, *Las colonias alemanas de Sierra Morena*, Doctoral thesis, Madrid, 1930.
46 Gonzalo Anes, 'El Antiguo Régimen: los Borbones', in *Historia de España Alfaguara*, vol. IV. Madrid: Alianza Editorial, 1983, pp. 142–3. On the Mesta, see the Wikipedia article 'Honrado Concejo de la Mesta'.

V. Free Migration

1 José Saramago, *Memoriale del convento*. Milan: Feltrinelli, 1984, p. 198.
2 António Filipe Pimentel, *Arquitectura e poder. O Real edifício de Mafra*. Lisbon: Livros Horizonte, 1992. The construction labour force reached a maximum of 45,000 in 1730, housed in a village-camp. Seven thousand soldiers oversaw the entire project. The annual average number of construction workers was around 15,000.
3 António de Oliveira, 'Migrações internas e de média distância em Portugal de 1500 a 1900', in Antonio Eiras Roel and Ofelia Rey Castelao (eds.), *Migraciones internas y média distance en la Península Ibérica, 1500–1900*, vol. II. Santiago de Compostela:

Xunta de Galicia y Comité international des sciences historiques, 1994.

4 Antonio Eiras Roel, 'Migraciones internas y medium distance en España en la Edad Moderna', in Eiras Roel and Rey Castelao (eds.), *Migraciones*, p. 40.

5 Ibid.

6 Oliveira, 'Migrações', p. 11.

7 Galanti, quoted by Ercole Sori, *L'emigrazione italiana dall'unità alla seconda guerra mondiale*. Bologna: il Mulino, 1979, p. 14.

8 Carlo Corsini, 'Le migrazioni stagionali di lavoratori nei dipartimenti italiani durante il periodo napoleonico (1810–12)', in Marco Breschi and Lorenzo Del Panta (eds), *Carlo Corsini: saggi di vita*. Udine: Forum, 2018.

9 Gérard Delille, 'Migrations paysannes et migrations des élites en Italie du Sud pendant la période moderne', in Eiras Roel and Rey Castelao (eds.), *Migraciones*, vol. I, p. 344.

10 On the Mesta, see above, note 46 of chapter IV. Eighteenth-century Spain reached a height of five million transhumant sheep, in flocks that generally did not exceed 100 head each. The number of shepherds involved in transhumance must therefore have amounted to several tens of thousands.

11 Dudley Baines, 'Internal and Medium-Distance Migrations in Great Britain 1500–1750', in Eiras Roel and Rey Castelao (eds), *Migraciones*, vol. I, p. 131.

12 Sune Akerman, 'Time of the Great Mobility: The Case of Northern Europe', in Eiras Roel and Rey Castelao (eds), *Migraciones*, vol. I, p. 82.

13 Jan Lucassen, *Migrant Labour in Europe, 1600–1900*. London: Croom Helm, 1987, p. 6.

14 Ibid., pp. 28–9, 100, 146–53.

15 Steve Hochstadt, 'Migration in Preindustrial Germany', *Central European History*, 16(3), 1983, pp. 211–12.

16 Cited in ibid.

17 An Irish ballad, cited in William F. Adams, *Ireland and Irish Emigration to the New World. From 1815 to the Famine*. New

Haven, CT: Yale University Press, 1932, pp. 207–8. 'Columbia' here refers to 'America'; O'Connell was an Irish patriot.
18 Paola Corti, *Storia delle migrazioni internazionali*. Bari: Laterza, 2003, p. 34.
19 C.E. Snow, 'Emigration from Great Britain', in Willcox (ed.), *International Migration. II: Interpretations*, p. 246.
20 Adolph Jensen, 'Migration Statistics of Denmark, Norway and Sweden', ibid., pp. 283–312.
21 Felix Klezl, 'Austria', ibid., pp. 390–410.
22 Johnson, *A History*, p. 146.
23 For a historical and statistical review of European emigration, see Willcox's aforementioned volume; see also Maurice R. Davie, *World Immigration*. New York: Macmillan, 1936.
24 See the report by deputies Luzzatti and Pantano on the emigration bill approved on 31 January 1901, reported in *Manuale dell'emigrazione*. Florence: Barbèra, 1901, together with the text of the law and other documentation.
25 Blanca Sánchez-Alonso, 'Labour and Immigration', in Victor Bulmer-Thomas et al. (eds), *The Cambridge Economic History of Latin America*. Cambridge: Cambridge University Press, 2006, vol. II, p. 385.
26 It should be noted that the space of free movement in Europe has shrunk with the United Kingdom's exit from the EU. The COVID-19 pandemic also led to the closing of national borders, allowing countries to depart from the principles of free movement in order to limit the speed of contagion. It is to be hoped that similar derogations will not be requested or granted for other reasons that have nothing to do with the public health emergency.
27 Sinclair Lewis, *Arrowsmith*, cited by Davie, *World Immigration*, p. 187.
28 The 1790 census counted about 3.9 million inhabitants, of whom 3.2 million were white and 0.7 million black. The whites were 77 per cent from Britain, with another 4.4 per cent from Ireland; 7.4 per cent were of German origin and 3.3 per cent Dutch origin.

See Davie, *World Immigration*, p. 44. According to Fenske, *International Migration*, Germans in 1790 made up 8.4 per cent of the white population.

29 The study based on the 1920 census attributed 41.4 per cent of the population to descent from Great Britain and Northern Ireland; followed, in order, by Germany with 16.3 per cent; Ireland with 11.2 per cent; Poland with 4.1 per cent; and Italy with 3.6 per cent. See *Immigration Quotas on the Basis of National Origin*, 70th Congress, 2nd session, Senate Document 259, p. 5.

30 These pages refer to the characteristics of the advancing wave of European immigration to North America. They are not a summary of the overall migratory phenomenon, which was, as is well known, very complex, starting with the arrival of slaves from Africa and the subsequent fate of their descendants, who in 1790 already made up just under one-fifth of the US population. Nor do these pages deal with the great African-American emigration from the South to the urban centres of the north and west, from the 1910s up to the 1960s, or the various mass migratory movements of other ethnic groups, from Chinese coolies to Latinos, or Asians of various origins.

31 The 'Five Civilized Nations' were the Seminole, Cherokee, Chickasaw, Choctaw and Muscogee peoples (as well as their black slaves) who had forms of self-government. This sort of forced migration, justified by dubious treaties, took place in the 1830s and 1840s: see the Wikipedia entry for 'Indian removal'.

32 Considerations drawn from Massimo Livi-Bacci, *A Short History of Migration*. Cambridge: Polity, 2012, pp. 3–4.

33 Ammerman and Cavalli-Sforza, *La transizione neolitica*, pp. 82–3.

34 Albert J. Ammerman and Luigi L. Cavalli-Sforza, 'Measuring the Rate of Spread of Early Farming in Europe', *Man*, 6(4), 1971, pp. 674–88.

35 Luigi L. Cavalli-Sforza et al., *Storia e geografia dei geni umani*. Milan: Adelphi, 1997, p. 208.

36 Ray A. Billington, *The Westward Movement in the United States*. Princeton, NJ: Van Nostrand, 1959, p. 33.
37 Ibid., p. 35.
38 Ibid., p. 39.
39 The origin of this saying is disputed, although it has been attributed to Horace Greely, founder of the *New York Daily Tribune* and a politically influential figure.
40 Billington, *The Westward Movement*, p. 85.
41 It was Census Bureau director Francis A. Walker who proclaimed, based on the 1890 census, the 'closing' of the frontier, meaning that there was no longer any deserted, or rather semi-deserted territory (with a density of less than two inhabitants per square mile, i.e. per 2.59 square kilometres), ahead of the wave.
42 Michael R. Haines, 'The White Population of the United States, 1790–1920', in Haines and Steckel (eds.), *A Population History*, p. 324.
43 Ray A. Billington, *Westward Expansion: A History of the American Frontier*. New York: Macmillan, 1967, p. 10.
44 Turner first published this thesis in 1893, and then revised and incorporated it into other framings. See Frederick J. Turner, *The Frontier in American History*. New York: Holt, 1920.
45 Billington, *Westward Expansion*, pp. 88–9.
46 Marcus L. Hansen, *The Immigrant in American History*. Cambridge, MA: Harvard University Press, 1940, p. 61.
47 Ibid.

Reconsiderations
1 The concept of reproductive 'fitness' is elaborated in chapter 2 of Livi-Bacci, *A Short History*. But here the term migrant 'fitness' is used in the broadest sense, i.e. as the ability to fit profitably into a society other than the one of origin.
2 We also note the increasing frequency of forced migrations, as mentioned in the introduction to chapter II.

Index

Page numbers in *italics* denote a map/illustration

Adolf of Schauenburg, Count 86
Adrianople, Battle of (378 CE) 26
Africans 129
  and slave trade 4, 28–9, 130
agriculture 92, 117
  and 'demic spread' theory 122
  Great Plains 61
  prehistoric 121–3
  Spanish 101
  and transoceanic migration 117
Alamanni 23
America *see* United States
Americas 4
  European migration to 130
  map of migratory movements within *6*
  and slave trade 29–30
Andalusia 101–2
antiquity 8–27
apoikoi
  emigration of 12–13, 131
Argentina 106
  European emigration to 117
  immigration legislation 116

Italian migrants/migration 131, *plate 19*
Arkansas 62
Armenians
  removal of from Ottoman Empire and massacre of 30, 42, 43–4, 46
Atatürk, Mustafa Kemal 42, 44
Athens 15
  number of inhabitants 15–16
Augustus, Emperor
  Res gestae 16–18, *plate 3*
Australia 80, 119
Austria
  and emigration 115

Balkan War, First 40, 41, 44
Bantu peoples 122
Barbero, Alessandro 23–4
Belgium
  recruitment of mine workers 80
biodemographic perspective 129–30
blight, potato *see* Ireland
Bolívar, Simón 38

Index

Bosnia-Herzegovina 41
Brazil 106
  European emigration to 117
  great drought (Grande Seca)
    and emigration of *retirantes*
    (1877–9) 56, 58–60
  Haitian diaspora to after
    earthquake 66–7
  and immigration 116
  smallpox epidemic (1878) 60
Britain
  emigration in nineteenth century
    114
  immigration to New England 81
  Passenger's Acts 75
  seasonal migration 111

California 3, 62, 123–4
Canada
  European immigration 116, 117
  provisions given to inhabitants of
    by France 83
  see also Quebec
Caribbean 63–70
  hurricanes 64
Carlos III, King 101
Catherine the Great 79, 128, *plate 14*
  manifesto on immigration 93, 94,
    95–6
Central America
  climatic turbulence caused by the
    El Niño cycle 68–9
  drought (2014–18) 68, 70
  hurricanes 68
  impact of climate on migration
    68–9
  transits of migrants attempting to
    get into US from 69
Central and Eastern Europe 112–13
  Germanization of 3
Champlain, Samuel de 81
Charles III 79
Chechens
  deportation of in Soviet Union 51

Chinese
  deportation of in Soviet Union 50
Cicero 15
Cieza de León, Pedro 33
cities
  migration to 104, 108
climate change 70
climate migration 56
climate refugees 56
'coffin ships' 74
Colbert 3, 78, 80–1, 82–3
Colonial Land and Emigration
  Department 114
Columbus 64
Committee for Union and Progress
  41
conflicts
  expulsions and flight caused by
    31
Constantius Chlorus 23

de la Vega, Garcilaso 34
'demic spread' theory 122
deportations 3, 28, 32
  of Armenians from Ottoman
    Empire 43, 46, 48, 50, 51–2, 54
  of ethnic minorities in Soviet
    Union 5, 31, 48, 49, 50–3
  of Germans from Norka *plate 8*
  map *plate 7*
  Roman Empire 24
  Second World War 119
  of slaves from Africa 28–9
displacement/displaced persons 4,
  22, 28, 31–2, 36, 40, 43, 49–50,
  66, 77, 103, 129
Drang nach Osten (drive to the
  East) 78, 86–92, 132, *plate 12*
  circumstances and characteristics
    90
  elites and organizations
    responsible for 87–8, 90, 92
  equipping of German immigrants
    90

Drang nach Osten (drive to the East) (*cont.*)
  expansion of and places colonized 88
  factors contributing to success of 92
  factors driving 89–90, 91
  and the *locator* (populator) 90–1
  main axes of 87
  numbers involved 88–9
drought 56, 58–63
  Central America (2014–18) 68, 70
  and decline of Maya civilization 56
  Dust Bowl (Great Plains of North America) 3, 56, 60–3
  Grande Seca (Brazil) 56, 58–60, 63
  Dust Bowl migrations 3, 56, 60–3, plate 9

earthquakes 57
El Niño cycle 57, 68–9
El Salvador 68
epidemics 56
Eta Hurricane 68
Euclides da Cunha 59
Europe 88
  increase in population 117
  intra-European migration 119
  loosening of restrictive emigration policies and reasons 114–16, 117
  map of migratory movements within 5
  see also transoceanic migration

filles du roi 3, 78, 80–8, 128, 132, plate 13
  arrangements made for arrival in Quebec 84
  journey made 85
  marriage on arrival 84
  migration of to Quebec 78, 80–8

  profile 84
  selection of 83–5, 132
  success of migration 129
First Nations *see* Native Americans
First World War 41–2, 43, 46
fitness 131–2
floods 57
forced displacement *see* displacement/displaced persons
forced migration 2, 3, 28–32, 54, 119, 129
  Inca Empire 3, 30, 32–9
  modern times 31
  and Native Americans 121
  Ottoman Empire 3, 30–1, 40–6
  Roman Empire 12, 19, 21, 22
  slave trade 4, 19
  Soviet Union 31, 46–54
  see also deportations
France
  and *filles du roi see filles du roi*
  population seen as cause of wealth 82
  population in seventeenth century 81
  provisions given to inhabitants of Canada 83
  seasonal migration 111
  views on immigration 82–3
Francis I 100
Franks 23–4
Frederick the Great 79
free migration 2, 3, 4, 103–27, 129, 133
  in antiquity 11
  definition 103–4
  ending of international 120
  great transoceanic migration 105, 113–20, 128
  and labour markets 104–5, 106–13
  North America 106, 120–7
  as a rare phenomenon 104

Index

Gela 14
Germania
  ethnicities of *plate 2*
Germans
  deportations of in Soviet Union 52–3
  emigration to America 94, 98, 126–7
  emigration to the Volga *see* Volga Germans
  migration of to Maremma 79, 100–1, 110, 111, 130
Germany
  depopulation due to Thirty Years' War 83–4
  Drang nach Osten (drive to the East) *see Drang nach Osten*
  and emigration 115
  and labour market migration 112
global warming 56, 70
globalization 117
Goths 12, 23–5, 26–7, 128
  battle between Romans and *plate 4*
Great Crash 119
Great Plains (North America)
  Dust Bowl 60–3
  land use patterns 61
great transoceanic migration *see* transoceanic migration
Greece, ancient 11–16, 19, 128, 131
  demographic profile and number of migrants in colonies 15
  expansion of empire 12
  foundation of colonies and organized migration 3, 11, 12–16, 19, 20, 78
  map of Mediterranean colonies *plate 1*
  number of colonies in Italy 15
  number of inhabitants 15–16
Greece/Greeks, modern
  deportation of from Ukraine during Second World War 51

population exchanges with Ottoman Empire 30, 42, 44–5, 46, *plate 7*
war against Turks 42
Guatemala 68

Habsburg Empire 94, 99–100
Haiti/Haitians 57, 64–7, 128
  closure of immigration to United States under Trump 67
  deforestation 64–5
  earthquake (2010) 66–7, 70, *plate 10*
  given Temporary Protected Status in United States 67
  hurricanes 65
  independence (1804) 64
  migrants' remittances 65
  migratory exodus from 65
  overpopulation 65
  poverty 65–6
  sugar-cane plantations 64
  water pollution 65
Helmold, Bosau
  Chronica Slavorum 86
Hermondurians 21–2
Hispaniola 64
Homestead Act (1862) 124
Honduras 68
Humboldt, Alexander von 29
Hungary
  and emigration 115
Huns 23
Hurricane Mitch (1998) 68
hurricanes 57
  Caribbean 64
  Central America 68
Hyblon, King 13–14

Illinois 125
Inca Empire 3, 30, 32–9, 128
  forced internal migration and *mitimaes* 33–6, 39
  incentives for the *mitimaes* 35

Inca Empire (*cont.*)
  Lupaca people 35–6
  Potosí silver mines and annual recruitment of Indians to work in 37–9, *plate 6*
  reasons and motivations for internal migration 33–4, 35, 39
  redistribution of dispersed population into larger groups/villages under Spanish rule 36–7, 39
  road network 32–3, *plate 5*
  Spanish conquest of 36
  vertical system of control 35–6
India, partition of (1947) 31
Indian Appropriation Act (1851) 121
Indian Removal Act (1830) 121
Indiana 123
Indians *see* Native Americans
Industrial Revolution 102
Ingrian Finns 50
Ingush, deportations of 51
Iota Hurricane 68
Ireland
  decimation of population by potato blight and Great Famine 74
  emigration to America due to Great Famine 71, 74–6, 119, *plate 17*
  impact of emigration on populations of origin 130
  marriage 72, 73, 76
  potato blight and Great Famine 57, 70–6, *plate 11*
  potato cultivation 72
  tripling of population before the Famine and reasons 71–2
Italian migrants
  and Argentina 131, *plate 19*
Italy
  and emigration 115

emigration and recruiting agents 118
Greek colonies in 12–15
seasonal migration 110–11

Jews 118, 195
Juan V, King of Portugal 107

Kansas 62
Kazakhstan 48, 49
Kemal, Mustafa *see* Atatürk
Kennedy, John F. 70
Khrushchev, Nikita 52

labour markets 104–5, 106–13
  building of great monasteries and cathedrals 107–8
  migration due to creation of 104–5
  North Sea 112
  seasonal migration 108–11
  success of migration 130–1
Lausanne, Treaty of 42, 44, 45
Lenin, Vladimir 53
Lincoln, Abraham 124
locatores (populators) 90–1
Lombards 27
Lorimer, Frank 48
Louis XIV, King 3, 78, 81, 82, 83
Lupaca people 35–6

Mafra convent (Portugal) 107–8
Malthus/Malthusian perspective 10, 57, 65, 102, 130
Marcellinus, Ammianus 23
Maremma
  migration of German settlers to 79, 100–1, 110, 111, 130
Maria Theresa 79, 100
Maya civilization
  drought and decline of 56
Megara Hyblaea (Sicily) 13–14, 15
Mehmet VI 40
Mennonite Germans 98

mercantilist doctrine 79, 82, 94, 102, 105, 114
Mexican labourers
 recruitment of by United States 80
Mexico 68, 123
 fault line between 'northern triangle' and 68
migration
 causes and factors 9–10
 difficulty in defining 1
 as inherent to humans 7–8, 9
 map 5
 post-First World War evaporation of freedom of 119
 renewal and turnover of societies 9, 10
 success of 129–31
mitimaes 3, 33, 35, 39
Montenegro 41

Native Americans 106
 forced confinement on reservations 121, 125
 and forced migration 121
 impact of 'advancing wave' of migration on 3, 121, 124–5, 126
 nature 55–76
 role played in migration 55–6
 see also drought; hurricanes
Naxos 13
Neanderthals 121
Neolithic revolution 122
Nicaragua 68
Nigeria, expulsions from (1983) 31
Norka
 deportation of Germans from *plate 8*
North America 128
 'advancing wave' of migration 3, 106, 120–7
 Dust Bowl migrations 3, 56, 60–3, *plate 9*
 effect of great migration on identity of American people 126
 European emigration to 116, 117, 120, 125, 131
 free choice of migrants 125–6
 and German immigrants 94, 98, 126–7
 and Homestead Act (1862) 124
 impact of 'advancing wave' of migration on Native Americans 3, 121, 124–5, 126
 pioneer farmers 125, 126
 profile of immigrants 124, 125
 Race to the West 124
 railways 124
 spread and extension of the settlement of 121
 see also United States
North Sea labour market 112
'northern triangle' 68

Ohio 123
oikistés 12–14
Oklahoma 62
Olavide, Pablo de 101
organized migration 2, 3–4, 77–102, 103, 129
 definition 77
 Drang nach Osten (drive to the East) case 78, 86–92, 132, *plate 12*
 failure of Tuscan Maremma initiative 79, 100–1, 110, 111, 130
 filles du roi case 3, 78, 80–8, 128, 132, *plate 13*
 and foundation of Greek colonies 3, 11, 12–16, 19, 20, 78
 and Habsburg Empire 99–100
 and new villages of Sierra Morena and Andalusia 101–2
 Roman Empire 11–12, 19–20

organized migration (*cont.*)
  Volga Germans 52–3, 79, 93–9, 129–30
Ottoman Empire 30–1, 40–6
  defeat and dissolution of 3, 40, 42, 44, 46
  deportation of Greeks and population exchanges 30, 42, 44–5, 46, *plate 7*
  dissatisfaction and opposition to 41
  ethnic and religious minorities in 42
  forced migrations 3, 30–1, 40–6
  motivations for forced migration 45–6
  nationalism 46
  removal of Armenians and genocide committed 30, 42, 43–4, 46
  Turkization of Anatolia 44, 45
Oviedo, Fernández de 64

Panegyric in Honour of Constantius 23–4
Paul the Deacon 27
People's Commissariat for Internal Affairs (NKVD) 48
Peru, ancient 30, 32–9 *see also* Inca Empire
Peter the Great 79
population
  increase in European 117
  Malthusian perspective 10, 57, 65, 102, 130
  Mercantilist doctrines 79, 82, 94, 102, 105, 114
population exchanges
  map *plate 7*
  and Ottoman Empire 30, 42, 45, 46
Portugal
  building of Mafra Convent 107
Potemkin 99

Potosí (Brazil), silver mines 37–9, *plate 6*
poverty
  escaping of through migration 65, 105, 128–9
prehistoric migration 121–3
Prussia 79, 94
  repopulation and colonization policies 94
Pueblo Indians 56–7

Quebec
  biological 'founders' 81–2
  comparison of demographic data with France 85–6
  growth of 84, 85
  immigration to 81
  migration of *filles du roi* to see *filles du roi*
  population 81

Raleigh, Sir Walter 72
refugees 4, 31–2
  flow of in Soviet Union during Second World War 53
Rohingya
  expulsion and deportation of 32
Roman Empire 11–12, 16–27, 128
  army as driving force of migratory processes 17–19
  Augustus's *Res gestae* 16–18, *plate 3*
  barbarian pressures on frontiers and attempt to contain 23–5
  fall of the Western 27
  and forced migration 12, 19, 21, 22
  Goth migration into Roman lands 25–7
  Goth raids and battles with Romans 26, *plate 4*
  important role of state in migration 19, 20–1
  limes 12, 20, 21, 22, 23

management and operation of
    the borders 20–2
organized migration 11–12, 19–20
relations between army and
    'barbarians' 20–2
resettlement of soldiers/veterans
    in new colonies 17–19
Romania 41
Rostovtzeff, Michael 18
Rousseff, Dilma 67
Russia
  Jewish emigration 118, 195
Russian Empire 93–9, 102, 128
    Catherine the Great's manifesto
      on immigration and opening
      up of borders 93, 94, 95–6
    continuation of colonization
      process 99
    extension of rule 99
    later restrictions on immigration
      96
    population 99
    repopulation and colonization
      policies 94
    and Volga Germans *see* Volga
      Germans
    see also Soviet Union
Russo-Turkish War (1768–74) 99

sailing ships 75
Saxons 23
Scandinavia
    and emigration 114–15
    seasonal migration 111
Scythians 23
seasonal migration/migrants
      108–11, 112–13, 128
    France 111
    Italy 110–11
    Spain 109–10
Second World War 3, 31, 98, 119
    evacuations in Soviet Union
      during 49–50
    selection of migrants 131–3

Seneca 7–10, 12
Serbia 41
Seven United Provinces
    (Netherlands) 112
Seven Years' War (1756–63) 94, 96
Sevres, Treaty of 42
Sicily
    history of settlements in 13–14
Sierra Morena 101–2
Silesia 89
slaves/slave trade 4, 28–30, 129,
      130
    deportation of from Africa 28–9
Solzhenitsyn, Aleksandr 47
South America 29, 57, 68, 118, 131
Soviet Union 46–54
    allowing of deportees to return to
      their homelands 52
    conditions of deportee labour
      camps/'special settlements' 51
    conditions for uprooting millions
      of people 47–8
    deportation of Germans 52–3
    deportation of Koreans 50–1
    deportations and forced
      migrations of ethnic minorities
      during the Second World War
      31, 48, 49, 50–3
    deportations of the kulaks 48, 50
    evacuations during Second
      World War 49–50
    flows of refugees during Second
      World War 53
    Gulag archipelago 47
    industrialization under Stalin 47,
      48
    internal migration movement
      statistics (1926–39) 48
    regulation of internal migration
      48–9
    see also Russian Empire
Spain
    and new villages of Sierra Morena
      and Andalusia 101–2

Spain (*cont.*)
  seasonal migration 109–10
  transhumanization of shepherds/
    sheep 111, *plate 15*
Stalin, Joseph 47, 48, 53
steamships
  and Irish migrants 75
Sugambrians 22
Syracuse 13, 15, 16
Syrian civil war 32

Taino 64
Talon, Jean 82
Teutonic Order 87–8
Texas 123
Thirty Years' War (1618–48) 83–4
Thucles 13
Thucydides 13
Tiberius 22
time scale
  and success of migration 131
Titanic *plate 18*
Toledo, Viceroy 36–7, 38, 39
  Memoria 37
Trail of Tears 3, 121
Transcontinental Railroad
    plate 16
transoceanic migration 105, 113–20, 128
  and agriculture 117
  characteristics 118–19
  clandestine in nature 115
  destination and origin countries 117
  dissolving of colonial model 113–14
  fall in costs and times of voyages 118
  growth of 116–17
  and increase in demographic growth 117
  and recruiting agents 118
  root causes and factors behind 105–6, 117–18

  selection mechanisms of
    destination countries 132
Trump, Donald 67
Turkey *see* Ottoman Empire
Turkish National Movement 42
Turner, Frederick Jackson 126
Tuscan Maremma initiative 100–1

Ukraine, Russia's attack on 32
UNHCR 31
United States
  emigration of Irish to due to
    Great Famine 71, 74–6, 119,
    plate 17
  ethnic composition 120
  German emigration to 94, 98, 126–7
  recruitment of Mexican
    labourers 80
  restrictions on migration
    119–20
  Temporary Protected Status
    (TPS) offered to Haitian
    migrants (2010) 67
  see also North America
Uruguay 116, 117

Valens, Emperor 25, 26
Volga ASSR 98
Volga Germans 52–3, 79, 93–9, 129–30
  and Catherine the Great's
    proclamation/manifesto on
    immigration 94–6
  deportation from Norka *plate 8*
  journey to the Volga 96–7
  land and agricultural equipment
    assigned to 97
  mortality rate 97
  numbers of 96
  organization and settlement of 97
  rights and privileges as stated
    in Catherine the Great's
    manifesto 95–6

and Second World War 98
subsequent history of 98
success of despite hardships 97–9, 129–30
voluntary migration *see* free migration

Werfel, Franz 44

Young Turk movement 41
Yugoslavian wars 31

Zosimus 25